COMETS AND METEORS

LECTOR HOUSE PUBLIC DOMAIN WORKS

COMETS AND METEORS

DANIEL KIRKWOOD

ISBN: 978-93-5614-222-0

Published: 1873

LECTOR HOUSE LLP
E-MAIL: lectorpublishing@gmail.com

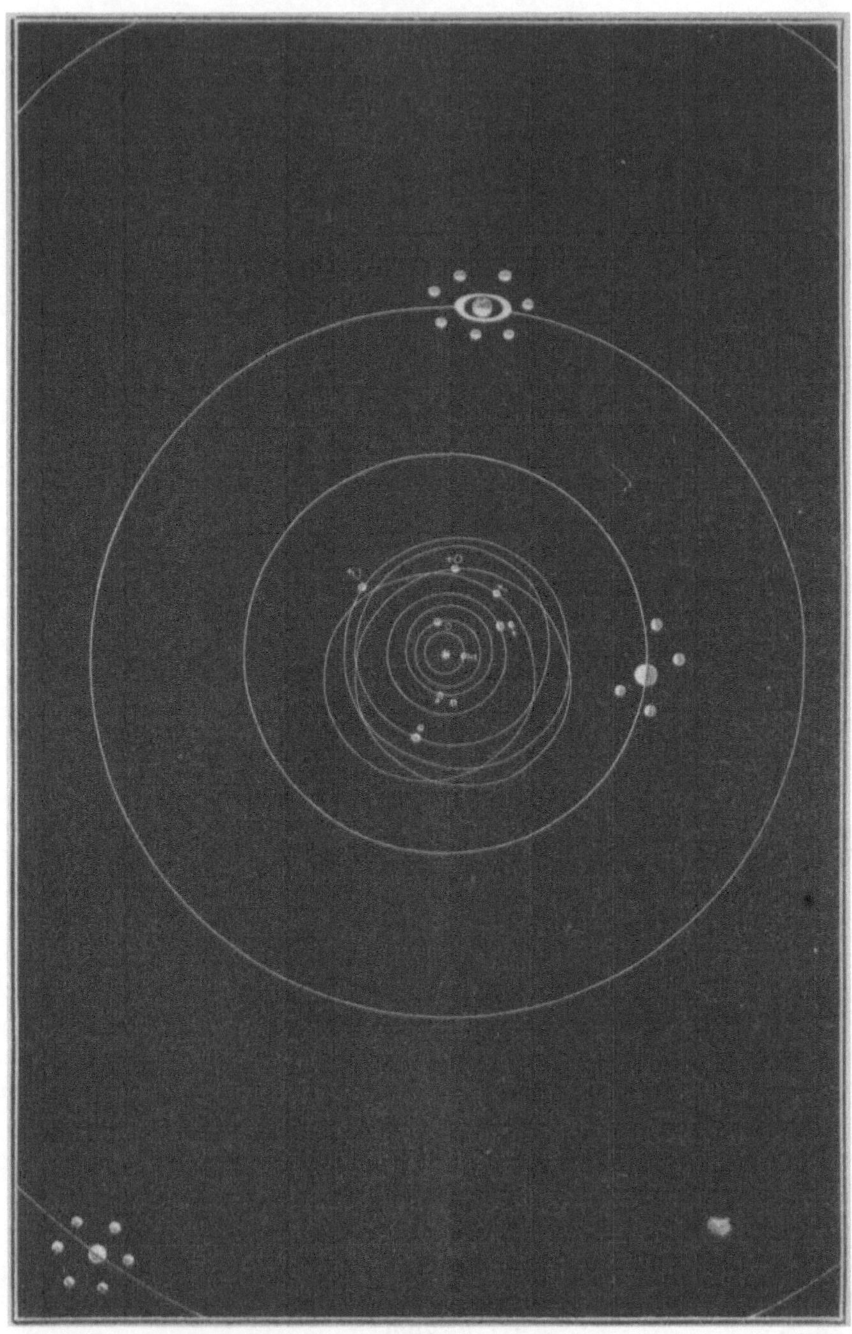

The Solar System.

COMETS AND METEORS:

THEIR PHENOMENA IN ALL AGES; THEIR MUTUAL RELATIONS; AND THE THEORY OF THEIR ORIGIN.

BY

DANIEL KIRKWOOD, LL.D.,

PROFESSOR OF MATHEMATICS IN INDIANA UNIVERSITY,
AND AUTHOR OF "METEORIC ASTRONOMY."

1873.

PREFACE.

The origin of meteoric astronomy, as a science, dates from the memorable star-shower of 1833. Soon after that brilliant display it was found that similar phenomena had been witnessed, at nearly regular intervals, in former times. This discovery led at once to another no less important, viz.: that the nebulous masses from which such showers are derived revolve about the sun in paths intersecting the earth's orbit. The theory that these meteor-clouds are but the scattered fragments of disintegrated comets was announced by several astronomers in 1867:—a theory confirmed in a remarkable manner by the shower of meteors from the *débris* of Biela's comet on the 27th of November, 1872.

To gratify the interest awakened in the public mind by the discoveries here named, is the main design of the following work. Among the subjects considered are, cometary astronomy; aerolites, with the phenomena attending their fall; the most brilliant star-showers of all ages; and the origin of comets, aerolites, and falling stars.

It may be proper to remark that the language used by the writer in a volume[1] published several years since, and now nearly out of print, has been occasionally adopted in the following treatise.

BLOOMINGTON, INDIANA, April, 1873.

---

[1] Meteoric Astronomy.

CONTENTS

I.
COMETS.

COMETS AND METEORS.

CHAPTER I.

A GENERAL VIEW OF THE SOLAR SYSTEM.

A descriptive treatise on Comets and Meteors may properly be preceded by a brief general view of the *planetary* system to which these bodies are related, and by which their motions, in direction and extent, are largely influenced.

The Solar System consists of the sun, together with the planets, comets, and meteors which revolve around it as the centre of their motions. The sun is the great controlling orb of this system, and the source of light and heat to its various members. Its magnitude is one million three hundred thousand times greater than that of the earth, and it contains more than seven hundred times as much matter as all the planets put together.

Mercury is the nearest planet to the sun; its mean distance being about 35,400,000 miles. Its diameter is 3000 miles, and it completes its orbital revolution in 88 days.

Venus, the next member of the system, is sometimes our morning and sometimes our evening star. Its magnitude is almost exactly the same as that of the earth. It revolves round the sun in 225 days.

The earth is the third planet from the sun in the order of distance; the radius of its orbit being about 92,000,000 miles. It is attended by one satellite, — the moon, — the diameter of which is 2160 miles.

Mars is the first planet exterior to the earth's orbit. It is considerably smaller than the earth, and has no satellite. It revolves round the sun in 687 days.

The Asteroids. —Since the commencement of the present century a remarkable zone of telescopic planets has been discovered immediately exterior to the orbit of Mars. These bodies are extremely small; some of them probably containing less matter than the largest mountains on the earth's surface. 131 members of the group are known at present, and the number is annually increasing.

Jupiter, the first planet exterior to the asteroids, is nearly 500,000,000 miles from the sun, and revolves round it in a little less than 12 years. This planet is 86,000 miles in diameter, and contains more than twice as much matter as all the other planets, primary and secondary, put together. Jupiter is attended by four moons or satellites.

Saturn is the sixth of the principal planets in the order of distance. Its orbit is about 400,000,000 miles beyond that of Jupiter. This planet is attended by eight satellites, and is surrounded by three broad flat rings. Saturn is 73,000 miles in diameter, and its mass or quantity of matter is more than that of all the other planets except Jupiter.

Uranus is at double the distance of Saturn, or nineteen times that of the earth. Its diameter is about 34,000 miles, and its period of revolution 84 years. It is attended by at least four satellites.

Neptune is the most remote known member of the system; its distance being 2,800,000,000 miles. It is somewhat larger than Uranus; has certainly one satellite, and probably several more. Its period is about 165 years. A cannon-ball flying outward from the sun at the uniform velocity of 500 miles per hour would not reach the orbit of Neptune in less than 639 years.

These planets all move round the sun in the same direction,—from west to east. Their motions are nearly circular, and also nearly in the same plane. Their orbits, except that of Neptune, are represented in the frontispiece. It is proper to remark, however, that all representations of the solar system by maps and planetariums must give an exceedingly erroneous view either of the magnitudes or distances of its various members. If the earth, for instance, be denoted by a ball half an inch in diameter, the diameter of the sun, according to the same scale (16,000 miles to the inch), will be between four and five feet; that of the earth's orbit, about 1000 feet; while that of Neptune's orbit will be nearly six miles. To give an accurate representation of the solar system at a single view is therefore plainly impracticable.

THE ZODIACAL LIGHT.—This term was first applied by Dominic Cassini, in 1683, to a faint nebulous aurora, somewhat resembling the milky way, apparently of a conical or lenticular form, having its base toward the sun and its axis nearly in the direction of the ecliptic. The most favorable time for observing it is when its axis is most nearly perpendicular to the horizon. This, in our latitudes, occurs in March, for the evening, and in October, for the morning. The angular distance of its vertex from the sun is frequently seventy or eighty degrees, while sometimes, though rarely (except within the tropics), it exceeds even one hundred degrees. It was noticed in the latter part of the 16th century by Tycho Brahe. The first accurate description of the phenomenon was given, however, by Cassini. This astronomer supposed the appearance to be produced by the blended light of innumerable bodies too small to be separately observed,—a theory still very generally accepted. In other words, the zodiacal light is probably a swarm of infinitesimal planets; the greater part of the cluster being interior to Mercury's orbit.

The distances between the different members of our planetary system, vast as they may seem, sink into insignificance when compared with the intervals which separate us from the so-called fixed stars. *Alpha Centauri*, the nearest of those twinkling luminaries, is 7000 times more distant than Neptune from the sun. Even light itself, which moves 185,000 miles in a second, is more than three years in traversing the mighty interval.

CHAPTER II.

COMETS.

The term *comet*—which signifies literally a *hairy star*—may be applied to all bodies that revolve about the sun in very eccentric orbits. The sudden appearance, vast dimensions, and extraordinary aspect of these celestial wanderers, together with their rapid and continually varying motions, have never failed to excite the attention and wonder of all observers. Nor is it surprising that in former times, when the nature of their orbits was wholly unknown, they should have been looked upon as omens of impending evil, or messengers of an angry Deity. Even now, although modern science has reduced their motions to the domain of law, determined approximately their orbits, and assigned in a number of instances their periods, the interest awakened by their appearance is in some respects still unabated.

The special points of dissimilarity between planets and comets are the following:—The former are dense, and, so far as we know, solid bodies; the latter are many thousand times rarer than the earth's atmosphere. The planets *all* move from west to east; many comets revolve in the opposite direction. The planetary orbits are but slightly inclined to the plane of the ecliptic; those of comets may have any inclination whatever. The planets are observed in all parts of their orbits; comets, only in those parts nearest the sun.

The larger comets are attended by a *tail*, or train of varying dimensions, extending generally in a direction opposite to that of the sun. The more condensed part, from which the tail proceeds, is called the *nucleus*; and the nebulous envelope immediately surrounding the nucleus is sometimes termed the *coma*. These different parts are seen in Fig. 2, which represents the great comet of 1811.

Zeno, Democritus, and other Greek philosophers held that comets were produced by the collection of several stars into clusters. Aristotle taught that they were formed by exhalations, which, rising from the earth's surface, ignited in the upper regions of the atmosphere. This hypothesis, through the great influence of its author, was generally received for almost two thousand years. Juster views, however, were entertained by the celebrated Seneca, who maintained that comets ought to be ranked among the permanent works of nature, and that their disappearance was not an extinction, but simply a passing beyond the reach of our vision. The observations of Tycho Brahe first established the fact that comets move through the planetary spaces far beyond the limits of our atmosphere. The illustrious Dane, however, supposed them to move in circular orbits. Kepler, on the other hand, was no less in error in considering their paths to be rectilinear. James Bernoulli supposed comets to be the satellites of a very remote planet, invisible on account of its great distance,—such satellites being seen only in the parts of their orbits nearest the earth. Still more extravagant was the hypothesis of Descartes, who held that they were originally fixed stars, which, having gradually lost their light, could no longer retain their positions, but were involved in the vortices of the neighboring stars, when such as were thus brought within the sphere of the sun's illuminating power again became visible.

Fig. 2.

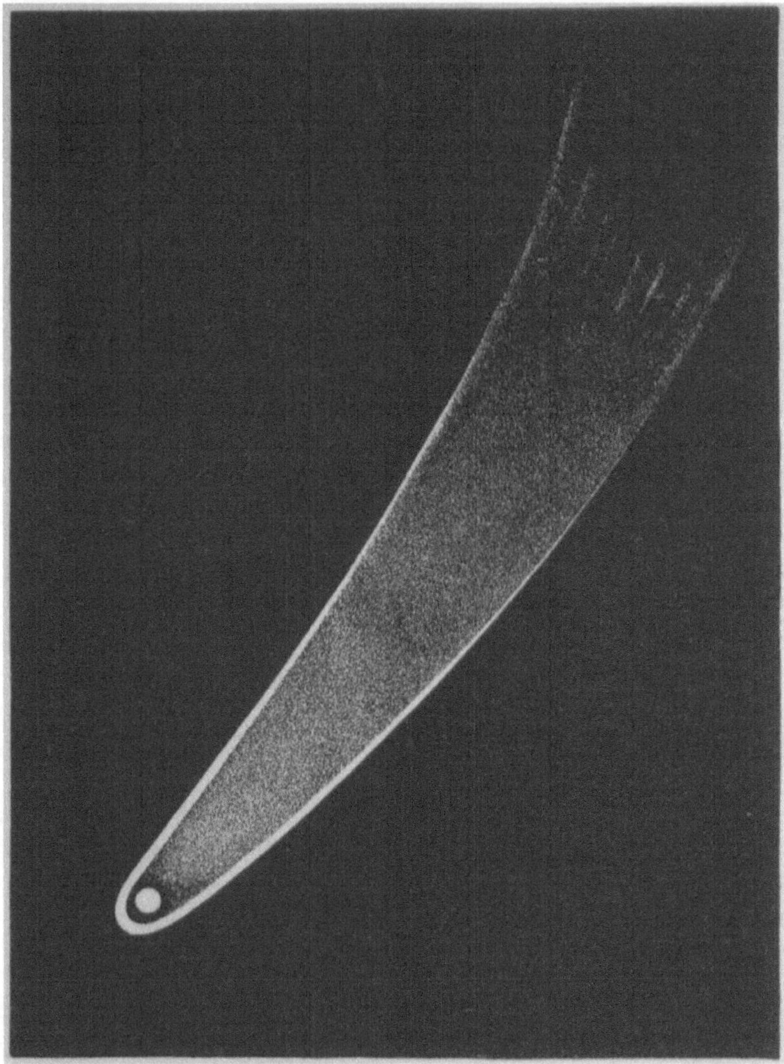

The Great Comet of 1811.

Comets visible in the Daytime.

Comets of extraordinary brilliancy have sometimes been seen during the daytime. At least thirteen authentic instances of this phenomenon have been recorded in history. The first was the comet which appeared about the year 43 B.C., just after the assassination of Julius Cæsar. The Romans called it the *Julium Sidus*, and regarded it as a celestial chariot sent to convey the soul of Cæsar to the skies. It was seen two or three hours before sunset, and continued visible for eight successive days. The great comet of 1106, described as an object of terrific splendor, was seen simultaneously with the sun, and in close proximity to it. Dr. Halley supposed this and the Julian comet to have been previous visits of the great comet of 1680. In the year 1402 two comets appeared,—one about the middle of February, the other in June,—both of which were visible while the sun was above the horizon. One was of such magnitude and brilliancy that the nucleus and even the tail could be seen at midday. The comet of 1472, one of the most splendid recorded in history, was visible in full daylight, when nearest the earth, on the 21st of January. This comet, according to Laugier, moves very nearly in the plane of the ecliptic, its inclination being less than two degrees. Its least distance from our globe was only 3,300,000 miles. The comet of 1532, supposed by some to be identical with that of 1661, was also visible in full sunshine. The apparent magnitude of its nucleus was three times greater than that of Jupiter. The comet of 1577 was seen with the naked eye by Tycho Brahe before sunset. It was by observations on this body that Aristotle's doctrine in regard to the origin, nature, and distance of comets was proved to be erroneous. It was simultaneously observed by Tycho at Oranienberg, and Thaddeus Hagecius at Prague; the points of observation being more than 400 miles apart, and nearly on the same meridian. The comet was found to have no sensible diurnal parallax; in other words, its apparent place in the heavens was the same to each observer, which could not have been the case had the comet been less distant than the moon. The comet which passed its perihelion on the 8th of November, 1618, was distinctly seen by Marsilius when the sun was above the horizon. The great comet of 1744 was seen without the aid of a glass at one o'clock in the afternoon,—only five hours after its perihelion passage. The diameter of this body was nearly equal to that of Jupiter. It had *six* tails, the greatest length of which was about 30,000,000 miles, or nearly one-third of the distance of the earth from the sun. The spaces between the tails were as dark as the rest of the heavens, while the tails themselves were bordered with a luminous edging of great beauty.

The great comet of 1843 was distinctly visible to the naked eye, at noon, on the 28th of February. It appeared as a brilliant body, within less than two degrees from the sun. This comet passed its perihelion on the 27th of February, at which time its distance from the sun's surface was only about one-fourth of the moon's distance from the earth. This is the nearest approach to the sun ever made by any known comet. The velocity of the body in perihelion was about 1,280,000 miles an hour, or nearly nineteen times that of the earth in its orbit. The apparent length of its tail was sixty-five degrees, and its true length 150,000,000 miles. The first comet of 1847, discovered by Mr. Hind, was also seen near the sun on the day of its perihelion passage. That discovered by Klinkerfues on the 10th of June, 1853, and which passed its perihelion on the 1st of September, was seen at Olmutz in the daytime, August 31, when only twelve degrees from the sun. After passing its perihelion, it was again observed, *at noon*, on the 2d, 3d, and 4th of September. Finally, the great comet of 1861 was seen before sunset, on Monday evening, July 1, by Rev. Henry W. Ballantine, of Bloomington, Indiana. It was again detected on the following evening just as the sun was in the horizon.

Besides the thirteen comets which we have enumerated, at least four others have been seen in the daytime; all, however, under peculiar circumstances. Seneca relates that during a great solar eclipse, 63 years before our era, a large comet was observed not far from the sun. "Philostorgius says that on the 19th of July, A.D. 418, when the sun was eclipsed and stars were visible, a great comet, in the form of a cone, was discovered near that luminary, and was afterwards observed during the nights."[2] The comet which passed its perihelion on the 18th of November, 1826, was observed by both Gambart and Flaugergues to transit the solar disk, — the least distance of the nucleus from the sun's surface being about 2,000,000 miles. The second comet of 1819 and the comet of 1823 are both known in like manner to have passed between the sun and the earth. Unfortunately, however, the transits were not observed.

A few cometary orbits are hyperbolas, more ellipses, and a still greater number parabolas. Comets moving in ellipses remain permanently within the limits of solar influence. Others, however, visit our system but once, and then pass off to wander indefinitely in the sidereal spaces.

Comets of known periodicity (Periodic Comets).

I. Halley's Comet.

As comets are subject to great changes of appearance, one can never be identified by any description of its magnitude, brilliancy, etc., at the time of a previous return. This can be done only by a comparison of orbits. If, for example, we find the elements of an orbit very nearly corresponding in every particular with those of a former comet, there is a degree of probability, amounting almost to certainty, that the two are identical. Sir Isaac Newton, in his *Principia*, published shortly after the appearance of the comet of 1682, explained how the periods of those mysterious visitors might thus be ascertained, thus directing the attention of astronomers to the subject. Dr. Halley soon after undertook a thorough discussion of all the recorded cometary observations within his reach. In the course of his investigations he discovered that the path of the comet observed by Kepler in 1607 coincided almost exactly with that of the one which passed its perihelion in 1682. Hence he concluded that they were the same. He found also that the comet of 1531, whose course had been particularly observed by Apian, moved in the same path. The interval between the consecutive appearances being nearly 76 years, Halley announced this as the time of the comet's revolution, and boldly predicted its return in 1758 or 1759. The law of universal gravitation had at this time just been discovered and announced. But although its application to the determination of

---

[2] Hind.

planetary and cometary perturbations had not been developed, Halley was well aware that the attractive influence of Jupiter and Saturn might accelerate or retard the motion of the comet, so as to produce a considerable variation in its period. During the interval from 1682 to 1759, the application of the higher mathematics to problems in physical astronomy had been studied with eminent success. The disturbing effect of the two large planets, Jupiter and Saturn, was computed with almost incredible labor by Clairaut, Lalande, and Madame Lepaute. The result as announced by Clairaut to the Academy of Sciences in November, 1758, was that the period must be 618 days longer than that immediately preceding, and that the comet accordingly would pass its perihelion about the 13th of April, 1759. It was stated, however, that, being pressed for want of time, they had neglected certain quantities which might somewhat affect the result. The comet, in fact, passed its perihelion in March, within less than a month of the predicted time. When it is considered that the attraction of the earth was not taken into the account, and that Uranus, whose influence must have been sensible, had not then been discovered, this must certainly be regarded as a remarkable approximation.

But during the next interval of 76 years the theory of planetary perturbations had been more perfectly developed. The masses of Jupiter and Saturn had been determined with greater accuracy, and Uranus had been added to the known members of the planetary system. A nearer approximation to the exact time of the comet's perihelion passage in 1835 was therefore to be expected. Prizes were offered by two of the learned societies of Europe—the Academy of Sciences at Turin, and the French Institute—for the most perfect discussion of its motions. That of the former was awarded to Damoiseau,—that of the latter to Pontecoulant. The times assigned by these distinguished mathematicians for the comet's perihelion passage were very nearly the same, and differed but a few days from the true time. Had the present received mass of Jupiter been used in the calculations, Pontecoulant, it is believed, would not have been in error as much as 24 hours. It may be proper to remark that, during the entire period from 1759 to 1835, the position of Neptune was such that it could produce no considerable effect on the motion of the comet.

This interesting object will again return about 1911.

The visit of 1531 was the earliest that Halley succeeded in determining with any degree of certainty. Peter Apian, by whom it was at that time observed, was the first European to ascertain the fact that, as a general thing, the tails of comets are turned from the sun.[3] To confirm this discovery, he carefully followed the body in its progress through the constellations. By means of his recorded observations Halley was enabled to identify this comet with that of 1607 and 1682. The great comet of 1456 he *conjectured* to be the same, from the date of its appearance. Pingré subsequently confirmed this suspicion by a careful examination of the few trustworthy records that could be collected from the writers of that period.

From the earlier descriptions of this comet we infer that its brilliancy is gradually diminishing. In 1456 its tail, which was slightly curved like a sword or sabre,

---

[3] The Chinese, however, as appears from Biot's researches, had observed the same fact 700 years earlier. See Humboldt's Cosmos, vol. iv. (Bohn's ed.), p. 544.

extended two-thirds of the distance from the horizon to the zenith. The appearance of such an object, in a grossly superstitious age, excited throughout Europe the utmost consternation. The Moslems had just taken Constantinople, and were threatening to advance westward into Europe. Pope Calixtus III., regarding the comet as confederate with the Turk, ordered prayers to be offered three times a day for deliverance from both. The alarm, however, was of short duration. Within ten days of its appearance the comet reached its perihelion. Receding from the sun, the sword-like form began to diminish in brilliancy and extent; and finally, to the great relief of Europe, it entirely disappeared.

The perihelion passage of 1456 was, until recently, the earliest known. It was shown by Laugier, however, in 1843, that among the notices of comets extracted by Edward Biot from the Chinese records, were observations of a body in 1378, which was undoubtedly the comet of Halley. Further researches among these annals enabled the same astronomer to recognize two ancient returns, one in 760, the other in 451. Still more recently the distinguished English astronomer, Mr. Hind, has traced back the returns to the year 11 B.C. He remarks, however, that previous to that epoch, "the Chinese descriptions of comets are too vague to aid us in tracing any more ancient appearances," and that "European writers of these remote times render us no assistance." Let us now inquire whether the comet had probably made any former approach to the sun in an orbit nearly identical with the present. It is well known that the modern period of this body is considerably less than the ancient. Thus, the mean period since A.D. 1456 has been 75.88 years; while from 11 B.C. to 1456 A.D. it was 77.27 years. In determining the approximate dates of former returns, the ancient period should evidently be employed. Now, it is a remarkable fact that of more than 70 comets,[4] or objects supposed to be comets, whose appearance was recorded during the six centuries immediately preceding the year 11 B.C., but one—that of 166 B.C.—was observed at a date corresponding nearly to that of a former return of Halley's comet. Of this object it is merely recorded that "a torch was seen in the heavens." Whether this was a comet or some other phenomenon, it is impossible to determine. But as the comet of Halley was more brilliant in ancient than in modern times, it seems highly improbable that seven *consecutive* returns of so conspicuous an object should have been unrecorded, especially as twelve comets per century[5] were observed during the same period. It would appear, therefore, that the perihelion passage of 11 B.C. was in fact the first ever made by the comet, or at least the first in an orbit nearly the same as the present.

The motion of Halley's comet is retrograde. The point of its nearest approach to the sun is situated within the orbit of Venus. Its greatest distance from the centre of the system is nearly twice that of Uranus, or 36 times that of the earth. The comet is, consequently, subject to great changes of temperature. When nearest the sun its light and heat are almost four times greater than the earth's; when most remote, they are 1200 times less. In the former position, the sun would appear much larger than to us; in the latter, his apparent diameter would not greatly exceed that of Jupiter, as viewed from the earth. It would be difficult to conjecture what the

[4] See the Catalogues of Chambers and Williams.
[5] The average number.

consequences might be, were our planet transported to either of these extremes of the cometary path. In the perihelion, the waters of the ocean would undoubtedly be reduced to a state of vapor; in the aphelion, they would be solidified by congelation.

II. Encke's Comet.

It was formerly supposed that all comets have their aphelia far beyond the limits of the planetary system. In 1818, however, a small comet was discovered by Pons, the orbit of which was subsequently found to be wholly interior to that of Jupiter. Its elements were presented by Bouvard, in 1819, to the Board of Longitude at Paris. The form and position of the orbit were immediately found to correspond with those of a comet observed by several astronomers in 1805. The different appearances were consequently regarded as returns of the same body. Its elliptic orbit was calculated by Encke, who found its period to be only about three years and four months. Its perihelion is within the orbit of Mercury; its aphelion, between the asteroids and the orbit of Jupiter.

Encke's comet is invisible to the naked eye, except in very favorable circumstances; it has no tail; its motion, like that of the planets, is from west to east; and its orbit is inclined about 13° to the ecliptic.

A comparison of the successive periods of this interesting object has led to the discovery that its time of revolution is gradually diminishing; a fact regarded by Encke and other astronomers as indicating the existence of an ethereal medium.

III. Biela's Comet.

The discovery of Encke's comet of short period was followed, in 1826, by that of another, whose revolution is completed in about six years and eight months. It was observed on the 27th of February, by M. Biela, an Austrian officer; accordingly it has since been known as *Biela's comet*. On computing its elements and comparing them with those of former comets, it was found to have been observed in 1772 and 1805. Damoiseau having calculated the dimensions of the comet's elliptic path and the time of its return, announced as the result of his computations the remarkable fact that the orbits of the earth and comet intersect each other, and that the comet would cross the earth's path on the 29th of October, 1832. This produced no little alarm among the uneducated, especially in France. Even some journalists are said to have predicted the destruction of our globe by a collision with the comet. When the latter, however, passed the point of intersection at the predicted time, the earth was at a distance of 50,000,000 miles.

At the return of 1845-6, Biela's comet exhibited a most remarkable appearance. Instead of a single comet, it appeared as two distinct bodies moving together side by side, at a distance from each other somewhat less than that of the moon from the earth. Astronomers, anxious to determine whether the cometary fragments had continued separate during an entire revolution, awaited the next return with no ordinary interest. The *two* bodies appeared at the predicted time (September, 1852); their distance apart having increased to 1,250,000 miles. In 1859 the comet, on account of its proximity to the sun, entirely escaped detection. At the return in 1865-6 the position of the object was quite favorable for observation, yet the search of astronomers was again unsuccessful. In 1872 the body escaped detection both in Europe and America. One fragment was seen, however, at Madras, India, on the mornings of the 2d and 3d of December,—several weeks after its perihelion passage. The comet's non-appearance in 1866 and its greatly diminished magnitude in 1872 leave no room to doubt its progressive dissolution. This subject will again be referred to in discussing the phenomena of meteoric showers.

IV. Faye's Comet.

On the 22d of November, 1843, M. Faye, of the Paris Observatory, discovered a comet, which was shown by Dr. Goldschmidt to revolve in an elliptic orbit, the perihelion of which is exterior to the orbit of Mars, and the aphelion immediately beyond that of Jupiter. The eccentricity is, therefore, less than that of any other comet previously discovered. Its period is about 7 years and 5 months.

It is possible that a comet moving in a parabola or hyperbola, with the sun in the focus, may be thrown into an elliptic orbit by the disturbing influence of Jupiter or one of the other large planets. The celebrated Leverrier undertook to determine whether the comet of Faye had in this manner been recently fixed as a permanent member of the solar system. He found that it could not have been so introduced since 1747, and, consequently, that it must have completed at least thirteen revolutions before its discovery.

This comet has been observed at each return from 1843 to the present time.

V. De Vico's Comet.

On the 22d of August, 1844, De Vico, of Rome, discovered a comet whose orbit is included between those of the earth and Jupiter. Its period is 1996 days, or about 5½ years. This body, from some cause,—perhaps a gradual dissolution,—has not been observed at any subsequent return.

VI. Brorsen's Comet.

On the 26th of February, 1846, Mr. Brorsen, of Kiel, discovered a faint comet, the mean distance and period of which are almost identical with those of De Vico's. This comet was not observed during the perihelion passage of 1851, on account of its unfavorable position with respect to the sun. It has, however, been subsequently detected.

VII. D'Arrest's Comet.

Dr. D'Arrest discovered a comet on the 27th of June, 1851, which was soon found to move in an elliptic orbit, with a period of about 6½ years. It entirely escaped observation, both in Europe and America, during its perihelion passage in 1857. It was observed, however, at the Cape of Good Hope. Its invisibility in 1864 was due to its unfavorable position. At its return in 1870, it was first seen on the

31st of August, by Dr. Winnecke, of Carlsruhe.

VIII. Tuttle's Comet.

A faint telescopic comet was discovered at the Observatory of Harvard College, on the evening of January 4, 1858, by Mr. H. P. Tuttle. The same body was independently found one week later by Dr. Bruhns, of Berlin. From observations made at Cambridge, Massachusetts, and Ann Arbor, Michigan, its elements were soon computed by different astronomers; the result in each case coinciding so closely with the elements of the second comet of 1790, as to place its identity wholly beyond doubt. Its period is nearly 13 years and 8 months. It had returned, therefore, without detection, in the years 1803, 1817, 1831, and 1844. On its approach to perihelion in 1871, it was first detected by M. Borelly, of Marseilles.

IX. Winnecke's Comet.

The second comet of 1858 was discovered on the 8th of March, by Dr. Winnecke, of Bonn. This proved to be identical with the third comet of 1819, whose period was computed by Encke to be about 5½ years. It had therefore returned unperceived no less than six times between 1819 and 1858. At its return in 1863 it again escaped detection. The perihelion passage of 1869 was made on the 30th of June. The comet was seen as early as April 13, and, after passing the sun, as late as October 11. "Schönfeld states that in part of April and May it appeared to have not one, but several, centres of condensation, and Vogel says that, in the beginning of June, it had a much greater resemblance to a star-cluster than to a nebula." This phenomenon, it may be remarked, bore a striking resemblance to the appearances observed in the comets of 389, 1618, and 1661.

X. Tempel's Comet.

On the 19th of December, 1865, M. Tempel, of Marseilles, discovered a small comet, which continued visible four weeks, passing its perihelion January 11, 1866. Dr. Oppolzer, of Vienna, after a careful determination of its elements, announced the interesting fact that its orbit very nearly intersects those of the earth and Uranus; the perihelion being situated immediately within the former, and the aphelion a short distance exterior to the latter. The period, according to the same astronomer, is 33 years and 65 days. The identity of this comet with that of 1366 was suggested by Professor H. A. Newton soon after its appearance,—a suggestion which subsequent research has strongly corroborated. It is also highly probable that the comet observed in China, September 29, 1133, was a former return of the same body. In 1366 it was conspicuous to the naked eye, while in 1866 it was wholly invisible without a telescope,—a fact indicative of its gradual dissolution. The connection of this comet with the meteors of November 14 will be elsewhere considered.

XI. The Second Comet of 1867.

Another comet of short period was discovered by M. Tempel on the 3d of

April, 1867. Its orbit is the least eccentric of all known comets: the perihelion being exterior to the orbit of Mars; the aphelion interior to that of Jupiter. Its motion is direct, and it completes a revolution in 5 years and 8 months.

CHAPTER III.

COMETS WHOSE ELEMENTS INDICATE PERIODICITY, BUT WHOSE RETURNS HAVE NOT BEEN RECOGNIZED.

I. The Group whose periods are nearly equal to that of Uranus.

Since the commencement of the present century five comets have been discovered, which form, with Halley's, an interesting and remarkable group. The first of these was detected by Pons, on the 20th of July, 1812; the second by Olbers, on the 6th of March, 1815; the third by De Vico, on the 28th of February, 1846; the fourth by Brorsen, on the 20th of July, 1847; and the last by Westphal, on the 27th of June, 1852. The periods of these bodies are all nearly equal, ranging from 68 to 76 years; their eccentricities are not greatly different; the motions of all, except that of Halley's, are direct; and the distances of their aphelia are somewhat greater than Neptune's distance from the sun. Of this group, the comets of 1812 and 1846 seem worthy of special notice. The former became visible to the naked eye shortly after its discovery, and each continued visible about ten weeks. Their elements are as follows:

Perihelion Passage.	Long. of Perih'n.	Long. of A. Node.	Incl.	Peri'n Dist.	Eccen-tricity.	Peri-od.	Direc-tion.	Com-puter.
1812, Sept. 15*d.* 7*h.*	92° 51′	253° 33′	73° 57′	0.7771	0.94454	70.68*y*	D	Encke.
1846, Mar. 5*d.* 12*h.*	90° 31′	77° 37′	85° 6′	0.6637	0.96224	73.715	D	Peirce.

The wonderful similarity of these elements, except in the longitude of the ascending node, is at once apparent. It will also be noticed that the longitude of the *descending* node of the latter is very nearly coincident with that of the *ascending* node of the former. These remarkable coincidences are presented to the eye in the following diagram, where the dotted ellipse represents the orbit of the comet of 1812, and the continuous curve that of the comet of 1846.

It is infinitely improbable that these coincidences should be accidental; they point undoubtedly to a common origin of the two bodies.

Fig. 3.

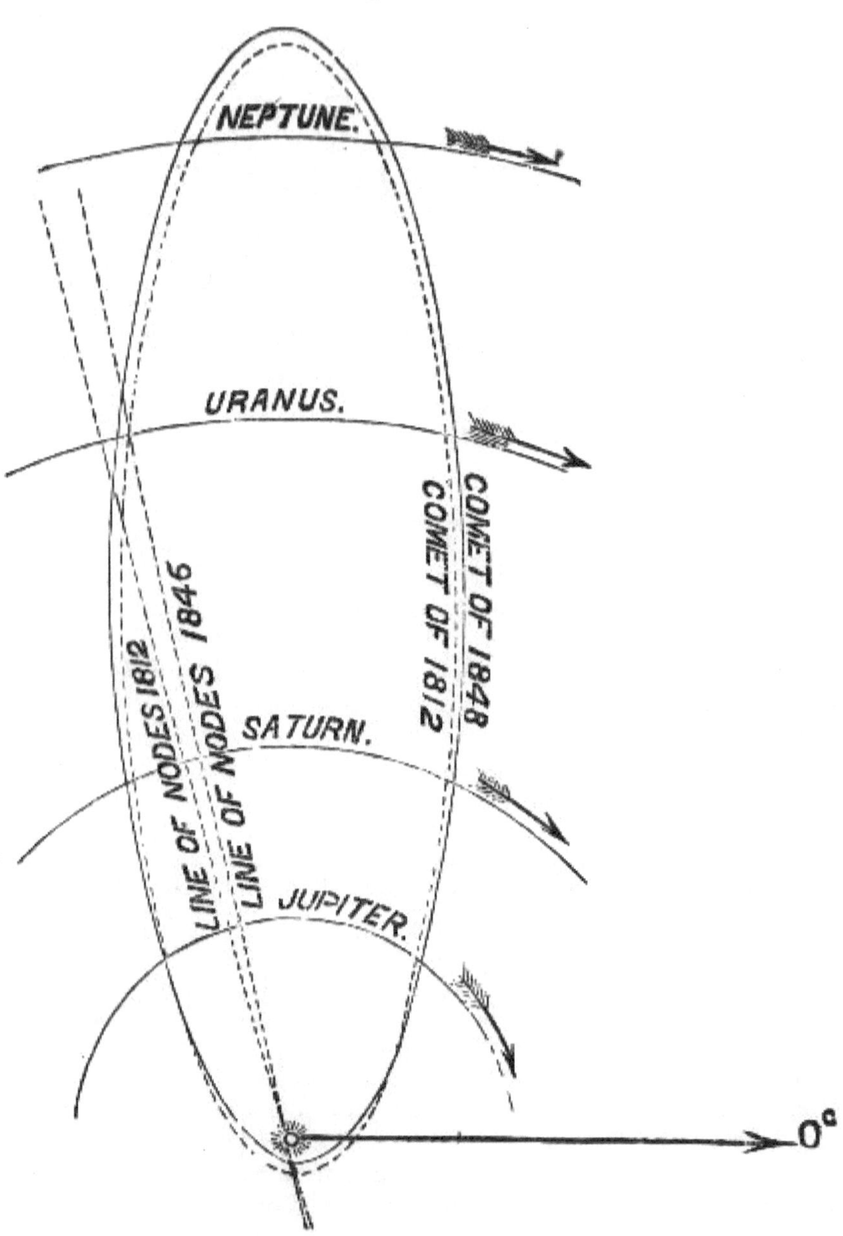

According to the theory now generally accepted, comets enter the solar system *ab extra*, move in parabolas or hyperbolas around the sun, and, if undisturbed by the planets, pass off beyond the limits of the sun's attraction, to be seen no more. If in their motion, however, they approach very near any of the larger planets, their direction is changed by planetary perturbation,—their orbits being sometimes transformed into ellipses. The new orbits of such bodies would pass very nearly through the points at which their greatest perturbation occurred; and accordingly we find that the aphelia of a large proportion of the periodic comets are near the orbits of the major planets. "I admit," says M. Hoek, "that the orbits of comets are by nature parabolas or hyperbolas, and that in the cases when elliptical orbits are met with, these are occasioned by planetary attractions, or derive their character from the uncertainty of our observations. To allow the contrary would be to admit some comets as permanent members of our planetary system, to which they ought to have belonged since its origin, and so to assert the simultaneous birth of that system and of these comets. As for me, I attribute to these a primitive wandering character. Traveling through space, they move from one star to another in order to leave it again, provided they do not meet any obstacle that may force them to remain in its vicinity. Such an obstacle was Jupiter, in the neighborhood of our sun, for the comets of Lexell and Brorsen, and probably for the greater part of periodical comets; the other part of which may be indebted for their elliptical orbits to the attractions of Saturn and the remaining planets.

"Generally, then, comets come to us from some star or other. The attraction of our sun modifies their orbit, as had been done already by each star through whose sphere of attraction they had passed. We can put the question if they come as single bodies or united in systems."

The conclusion of this astronomer's interesting discussion is that—

"There are systems of comets in space that are broken up by the attraction of our sun, and whose members attain, as isolated bodies, the vicinity of the earth during a course of several years." [6]

In the researches here referred to, it is shown by Professor Hoek that the comets of 1860 III., 1863 I., and 1863 IV. formed a group in space previous to their entrance into our system. The same fact has also been demonstrated in regard to other comets which need not here be specified. Now, the comets of 1812 and 1846 IV. have their aphelia near the orbit of Neptune, and hence the original parabolas in which they moved were probably transformed into ellipses by the perturbations of that planet. Before entering the solar domain, they were doubtless members of a cometary system. Passing Neptune near the same time, and at some distance from each other, their different relative positions with regard to the disturbing body may account for the slight differences in the elements of their orbits.

Comets of the Jovian Group.

Besides the eight comets enumerated in Chapter II. whose aphelia are in the vicinity of Jupiter's orbit, five others have been observed which belong apparently to the same cluster. These are the comets of 1585, 1743 I., 1766 II., 1783, and 1819 IV. "The fact that these comets have not been re-observed on their successive returns through perihelion may be explained either by the difficulty of observing them, owing to their unfavorable positions, and to the circumstances of observers not expecting their reappearance, their periodic character not being then suspected, or because they may have been thrown by the disturbing action of the larger planets

[6] Monthly Notices of the R. A. S., vol. xxv., p. 243.

into orbits such as to keep them continually out of the range of view of terrestrial observers."[7]

Lexell's comet of 1770 is the most remarkable instance known of the change produced in the orbits of these bodies by planetary attraction. This comet passed so near Jupiter in 1779 that the attraction of the latter was 200 times greater than that of the sun. The consequence was that the comet, whose mean distance corresponded to a period of 5½ years, was thrown into an orbit so entirely different that it has never since been visible.

PETERS' COMET.

A telescopic comet was discovered by Dr. Peters on the 26th of June, 1846, which continued to be observed till the 21st of July. Its period, according to the discoverer, is about 13 years, and its aphelion, like that of Tuttle's comet, is in the vicinity of Saturn's orbit. It was expected to return in 1859, and again in 1872, but each time escaped detection, owing probably to the fact that its position was unfavorable for observation.

STEPHAN'S COMET (1867 I.).

In January, 1867, M. Stephan, of Marseilles, discovered a new comet, the elements of which, after two months' observations, were computed by Mr. G. M. Searle, of Cambridge, Massachusetts. The perihelion of this body is near the orbit of Mars; its aphelion near that of Uranus,—the least distance of the orbits being about 2,000,000 miles. The present form of the cometary path is doubtless due to the disturbing action of Uranus. The comet completes its revolution in 33.62 years; consequently (as has been pointed out by Mr. J. R. Hind) five of its periods are almost exactly equal to two periods of Uranus. The next approximate appulse of the two bodies will occur in 1985, when the form of the comet's orbit may be sensibly modified.

ELLIPTIC COMETS WHOSE APHELIA ARE AT A MUCH GREATER DISTANCE THAN NEPTUNE'S ORBIT.

In October, 1097, a comet was seen both in Europe and China, which was noted for the fact of its having two distinct tails, making with each other an angle of about 40°. From a discussion of the Chinese observations (which extended through a longer period than the European), Laugier concluded that this body is identical with the third comet of 1840, which was discovered by Galle on the 6th of March. If, therefore, it has made no intermediate return without being observed, it must have a period of about 743 years. It is also highly probable, from the similarity of elements, that the comet which passed its perihelion on the 5th of June, 1845, was a reappearance of the comet of 1596,—the period of revolution being 249 years. The elements of the great comet of 1843 are somewhat uncertain. There is a probability, however, of the identity of this body with the comet of 1668. This would make the period 175 years. The third comet of 1862 is especially interesting from its connec-

[7] Dr. Lardner.

tion with the August meteors. Its period, according to Dr. Oppolzer, is 121½ years.

The Great Comet of 1858

was one of the most remarkable in the nineteenth century. It was discovered on the 2d of June, by Donati, of Florence, and first became visible to the naked eye about the last of August. The comet attained its greatest brilliancy about the 10th of October, when its distance from the earth was 50,000,000 miles. The length of its tail somewhat exceeded this distance. If, therefore, the comet had been at that time directly between the sun and the earth, the latter must have been enveloped for a number of hours in the cometic matter.

The observations of this comet during a period of five months enabled astronomers to determine the elements of its orbit within small limits of error. It completes a revolution, according to Newcomb, in 1854 years, in an orbit somewhat more eccentric than that of Halley's comet. It will not return before the 38th century, and will only reach its aphelion about the year 2800. Its motion per second when nearest the sun is 36 miles; when most remote, only 234 yards.

CHAPTER IV.

OTHER REMARKABLE COMETS.

It remains to describe some of the most remarkable comets of which we have any record, but of which we have no means of determining with certainty whether they move in ellipses, parabolas, or hyperbolas.

In the year 466 B.C., a large comet appeared simultaneously with the famous fall of meteoric stones near Ægospotamos. The former was supposed by the ancients to have had some agency in producing the latter phenomenon. Another of extraordinary magnitude appeared in the year 373 B.C. This comet was so bright as to throw shadows, and its tail extended one-third of the distance from the horizon to the zenith. The years 156, 136, 130, and 48, before our era, were also signalized by the appearance of very large comets. The apparent magnitude of the first of these is said to have equaled that of the sun itself; while its light was sufficient to diminish sensibly the darkness of the night. The second is said to have filled a fourth part of the celestial hemisphere. The comet of 130 B.C., sometimes called the comet of Mithridates, because of its appearance about the time of his birth, is said to have rivaled the sun in splendor.

In A.D. 178 a large comet was visible during a period of nearly three months. Its nucleus had a remarkably red or fiery appearance, and the greatest length of its tail exceeded 60°. The most brilliant comets of the sixth century were probably those of 531 and 582. The train of the latter, as seen in the west soon after sunset, presented the appearance of a distant conflagration.

Great comets appeared in the years 975, 1264, and 1556. Of these, the comet of 1264 had the greatest apparent magnitude. It was first seen early in July, and attained its greatest brilliancy in the latter part of August, when its tail was 100° in length. It disappeared on the 3d of October, about the time of the death of Pope Urban IV., of which event the comet, in consequence of this coincidence, was considered the precursor. These comets, on account of the similarity of their elements, were believed by many astronomers to be the same, and to have a period of about 300 years. In the case of identity, however, another reappearance should have occurred soon after the middle of the nineteenth century. As no such return was observed, we may conclude that the comets were not the same, and that their periods are wholly unknown.

The comet discovered on the 10th of November, 1618, was one of the largest in modern times; its tail having attained the extraordinary length of 104°. The comet of 1652, so carefully observed by Hevelius, almost equaled the moon in apparent magnitude. It shone, however, with a lurid, dismal light. The tail of the comet of 1680 was 90° in length. This body is also remarkable for its near approach to the sun; its least distance from the solar surface having been only 147,000 miles. It will always be especially memorable, however, for having furnished Newton the data by means of which he first showed that comets in their orbital motions are governed by the same principle that regulates the planetary revolutions.

Of all the comets which appeared during the eighteenth century, that which passed its perihelion on the 7th of October, 1769, had the greatest apparent magnitude. It was discovered by Messier on the 8th of August, and continued to be observed till the 1st of December. On the 11th of September the length of its tail was 97°. The comet discovered on the 26th of March, 1811, is in some respects the most remarkable on record. It was observed during a period of 16 months and 22 days,—the longest period of visibility known. On account of its situation with respect to the earth, the apparent length of its tail was much less than that of some other comets; its true length, however, was at one time 120,000,000 miles; and Sir William Herschel found that on the 12th of October the greatest circular section of the tail was 15,000,000 miles in diameter. The same astronomer found the diameter of the head of the comet to be 127,000 miles, and that of the envelope at least 643,000. As a general thing, the length of a comet-train increases very rapidly as the body approaches the sun. But the perihelion distance of the comet of 1811 was considerably greater than the distance of the earth from the sun; while its nearest approach to the earth was 110,000,000 miles. Its true magnitude, therefore, has probably not been surpassed by any other observed; and had its perihelion been very near the sun, it must have exhibited an appearance of terrific grandeur. This comet has an elliptic orbit, and its period, according to Argelander, is 3065 years.

The great comet of 1861 was discovered on the 13th of May, by Mr. John Tebbut, Jr., of New South Wales. In this country, as well as in Europe, it was first generally observed on the evening of June 30,—19 days after its perihelion passage. Sir John Herschel, who observed it in Kent, England, remarks that it far exceeded in brilliancy any comets he had ever seen, not excepting those of 1811 and 1858. According to Father Secchi, of the Collegio Romano, the length of its tail was 118°. This, with a single exception,[8] is the greatest on record. The computed orbit is elliptical; the period, 419 years.

---

[8] The tail of the first comet of 1865 (observed in the Southern Hemisphere) attained the unprecedented length of 150°.—*M. N. R. A. S.*, vol. xxv., p. 220.

CHAPTER V.

THE POSITION AND ARRANGEMENT
OF COMETARY ORBITS.

The cosmical masses from which comets are derived seem to traverse in great numbers the interstellar spaces. In consequence of the sun's progressive motion, these nebulous bodies are sometimes drawn toward the centre of our system. If, in this approach, they are not disturbed by any of the large planets, they again recede in parabolas or hyperbolas. When, however, as must sometimes be the case, they pass near Jupiter, Saturn, Uranus, or Neptune, their orbits may be transformed into elongated ellipses. The periodicity of many comets may thus be accounted for.

In the present chapter it is proposed to consider the probable consequences of the sun's motion through regions of space in which cometary matter is widely diffused; to compare our theoretical deductions with observed phenomena; and thus refer to their physical cause a variety of facts which have hitherto received no satisfactory explanation.[9]

1. As comets, at least in many instances, owe their periodicity to the disturbing action of the major planets, and as this planetary influence is sometimes sufficient, especially in the case of Jupiter and Saturn, to change the *direction* of cometary motion, the great majority of periodic comets should move in the same direction with the planets. Now, of the comets known to be elliptical, 70 per cent. *have direct motion*. In this respect, therefore, theory and observation are in striking harmony.

2. When the relative positions of a comet and the disturbing planet are such as to give the transformed orbit of the former a small perihelion distance, the comet must return to the point at which it received its greatest perturbation; in other words, to the orbit of the planet. The aphelia of the comets of short period ought therefore to be found, for the most part, *in the vicinity of the orbits of the major planets*. This, as already shown in Chapters II. and III., is strikingly the case. The actual distances of these aphelia, however, as compared with the respective distances of Jupiter, Saturn, Uranus and Neptune, are presented at one view in the following tables:

---

[9] This chapter is the substance of a paper read before the American Philosophical Society, November 19, 1869.

I. COMETS WHOSE APHELION DISTANCES ARE NEARLY EQUAL TO 5.20, THE RADIUS OF JUPITER'S ORBIT.

Comets.	Aph. Dist.
1. Encke's	4.09
2. 1819 IV	4.81
3. De Vico's	5.02
4. Pigott's (1783)	5.28
5. 1867 II	5.29
6. 1743 I	5.32
7. 1766 II	5.47
8. 1819 III	5.55
9. Brorsen's	5.64
10. D'Arrest's	5.75
11. Faye's	5.93
12. Bicla's	6.19

II. COMETS WHOSE APHELION DISTANCES ARE NEARLY EQUAL TO 9.54, THE RADIUS OF SATURN'S ORBIT.

Comets.	Aph.Dist.
1. Peters' (1846 VI.)	9.45
2. Tuttle's (1858 I.)	10.42

III. COMETS WHOSE APHELION DISTANCES ARE NEARLY EQUAL TO 19.18, THE RADIUS OF URANUS'S ORBIT.

Comets.	Aph. Dist.
1. 1867 I	19.28
2. November meteors	19.65
3. 1866 I	19.92

IV. COMETS WHOSE APHELION DISTANCES ARE NEARLY EQUAL TO 30.04, THE RADIUS OF NEPTUNE'S ORBIT.

Comets.	Aph. Dist.
1. Westphal's (1852 IV.)	31.97
2. Pons' (1812)	33.41
3. Olbers' (1815)	34.05
4. De Vico's (1846 IV.)	34.35
5. Brorsen's (1847 V.)	35.07

      6. Halley's[10]   35.37

The coincidences here pointed out (some of which have been noticed by others) appear, then, to be necessary consequences of the motion of the solar system through spaces occupied by meteoric nebulæ. Hence the observed facts receive an obvious explanation.

In regard to comets of long period we have only to remark that, for anything we know to the contrary, there may be causes of perturbation far exterior to the orbit of Neptune.

3. From what we observe in regard to the *larger* bodies of the universe—a clustering tendency being everywhere apparent,—it seems highly improbable that cometic matter should be uniformly distributed in the sidereal spaces. We would expect, on the contrary, to find it collected in groups or clusters. This view is also in remarkable harmony with the facts of observation. In 150 years, from 1600 to 1750, 16 comets were visible to the naked eye; of which 8 appeared in the 25 years from 1664 to 1689. Again, during 60 years, from 1750 to 1810, only 5 comets were visible to the naked eye, while in the next 50 years there were double that number. The probable cause of such variations is sufficiently obvious. As the sun in its progressive motion approaches a cometary group, the latter is drawn toward the centre of our system; the nearer members with greater velocity than the more remote. Those of the same cluster would enter the solar domain at periods not very distant from each other; the forms of their orbits depending upon their original relative positions with reference to the sun's course, and also on planetary perturbations. It is evident also that the passage of the solar system through a region of space comparatively destitute of cometic clusters would be indicated by a corresponding paucity of comets.

4. The line of apsides of a large proportion of comets will be approximately coincident with the solar orbit. The point towards which the sun is moving is in longitude about 260°. The quadrants bisected by this point and that directly opposite extend from 215° to 305°, and from 35° to 125°. The number of cometary perihelia found in these quadrants up to July, 1868 (periodic comets being counted but once) was 159, or 62 per cent.; in the other two quadrants, 98, or 38 per cent.

This tendency of the perihelia to crowd together in two opposite regions has been noticed by different writers.

5. Comets whose positions before entering our system were very remote from the solar orbit must have *overtaken* the sun in its progressive motion; hence their perihelia must fall, for the most part, in the vicinity of the point towards which the sun is moving; and they must in general have very small perihelion distances. Now, what are the observed facts in regard to the longitudes of the perihelia of the comets which have approached within the least distance of the sun's surface? But three have had a perihelion distance less than 0.01. *All* these, it will be seen by the following table, have their perihelia in close proximity to the point referred to:

---

[10] Halley's comet in aphelio is too remote from the plane of the ecliptic to be much disturbed by Neptune. Has the original position of the orbit been changed by Jupiter's influence?

I. COMETS WHOSE PERIHELION DISTANCES ARE LESS THAN 0.01.

Perihelion Passage.			Per. Dist.	Long. of Per.	
1. 1668, Feb.	28d.	13h.	0.0047	277°	2′
2. 1680, Dec.	17	23	0.0062	262	49
3. 1843, Feb.	27	9	0.0055	278	39

In Table II. all but the last have their perihelia in the same quadrant.

II. COMETS WHOSE PERIHELION DISTANCES ARE GREATER THAN
0.01 AND LESS THAN 0.05.

Perihelion Passage.			Per. Dist.	Long. of Per.	
1. 1689, Nov	29d.	4h.	0.0189	269°	41′
2. 1816, March	1	8	0.0485	267	35
3. 1826, Nov	18	9	0.0268	315	31
4. 1847, March	30	6	0.0425	276	2
5. 1865, Jan	14	7	0.0260	141	15

The perihelion of the first comet in Table III. is remote from the direction of the sun's motion; that of the second is distant but 14°, and of the third 21°.

III. COMETS WHOSE PERIHELION DISTANCES ARE GREATER
THAN 0.05 AND LESS THAN 0.1.

Perihelion Passage.			Per. Dist.	Long. of Per.	
1. 1593, July	18d.	13h.	0.0891	176°	19′
2. 1780, Sept.	30	22	0.0963	246	35
3. 1821, March	21	12	0.0918	239	29

With greater perihelion distances the tendency of the perihelia to crowd together round the point indicated is less distinctly marked.

6. Few comets of small perihelion distance should have their perihelia in the vicinity of longitude 80°, the point opposite that towards which the sun is moving. Accordingly we find, by examining a table of cometary elements, that with a perihelion distance less than 0.1 there is not a single perihelion between 35° and 125°; between 0.1 and 0.2 but 3; and between 0.2 and 0.3 only 1.

CHAPTER VI.

THE DISINTEGRATION OF COMETS.

The *fact* that in several instances meteoric streams move in orbits identical with those of certain comets was first established by the researches of Signor Schiaparelli. The *theory*, however, of an intimate relationship between comets and meteors was advocated by the writer as long since as 1861,[11]—several years previous to the publication of Schiaparelli's memoirs. In the essay here referred to it was maintained—

1. That meteors and meteoric rings "are the *débris* of ancient but now disintegrated comets whose matter has become distributed around their orbits."

2. That the separation of Biela's comet as it approached the sun in December, 1845, was but one in a series of similar processes which would probably continue until the individual fragments would become invisible.

3. That certain luminous meteors have entered the solar system from the interstellar spaces.[12]

4. That the orbits of some meteors and periodic comets have been transformed into ellipses by planetary perturbation; and

5. That numerous facts—some observed in ancient and some in modern times—have been decidedly indicative of cometary disintegration.

What was thus proposed as theory has been since confirmed as undoubted facts. When the hypothesis was originally advanced, the data required for its mathematical demonstration were entirely wanting. The evidence, however, by which it was sustained was sufficient to give it a high degree of probability.

The existence of a divellent force by which comets near their perihelia have been separated into parts is clearly shown by the following facts. Whether this force, as suggested by Schiaparelli, is simply the unequal attraction of the sun on different parts of the nebulous mass, or whether, in accordance with the views of other astronomers, it is to be regarded as a cosmical force of repulsion, is a question left for future discussion.

HISTORICAL FACTS.

1. Seneca informs us that Ephoras, a Greek writer of the fourth century be-

[11] Danville Quarterly Review, December, 1861.

[12] Others, it was supposed, might have originated within the system,—a view which the writer has not wholly abandoned.

fore Christ had recorded the singular fact of a comet's separation into two distinct parts.[13] This statement was deemed incredible by the Roman philosopher, inasmuch as the occurrence was then without a parallel. More recent observations of similar phenomena leave no room to question the historian's veracity.

2. The head of the great comet of A.D. 389, according to the writers of that period, was "composed of several small stars." (Hind's "Comets," p. 103.)

3. On June 27, A.D. 416, two comets appeared in the constellation Hercules, and pursued nearly the same apparent path. Probably at a former epoch the pair had constituted a single comet.[14]

4. On August 4, 813, "a comet was seen which resembled two moons joined together." They subsequently separated, the fragments assuming different forms.[15]

5. The Chinese annals record the appearance of three comets—one large and two smaller ones—at the same time, in the year 896 of our era. "They traveled together for three days. The little ones disappeared first, and then the large one."[16] The bodies were probably fragments of a large comet which, on approaching the sun, had been separated into parts a short time previous to the date of their discovery.

6. *The third comet of 1618.*—The great comet of 1618 exhibited decided symptoms of disintegration. When first observed (on November 30), its appearance was that of a lucid and nearly spherical mass. On the eighth day the process of division was distinctly noticed, and on the 20th of December it resembled a cluster of small stars.[17]

7. *The comet of 1661.*—The elements of the comets of 1532 and 1661 have a remarkable resemblance, and previous to the year 1790 astronomers regarded the bodies as identical. The similarity of the elements is seen at a glance in the following table:

	Comet of 1532.		Comet of 1661.	
Longitude of perihelion	111°	48′	115°	16′
Longitude of ascending node	87	23	81	54
Inclination	32	36	33	1
Perihelion distance	0.5192		0.4427	
Motion	Direct.		Direct.	

The elements of the former are by Olbers; those of the latter by Mechain. The return of the comet about 1790, though generally expected, was looked for in vain. As a possible explanation of this fact, it is interesting to recur to an almost forgotten statement of Hevelius. This astronomer observed in the comet of 1661 an

---

[13] "Quæst. Nat.," lib. vii., cap. xvi.
[14] Chambers' "Descr. Astr.," p. 374.
[15] Ibid., p. 383.
[16] Ibid., p. 388.
[17] Hevelius, "Cometographia," p. 341. See also Grant's "Hist. of Phys. Astr.," p. 302.

apparent breaking up of the body into separate fragments.[18] The case may be analogous to that of Biela's comet.

8. The identity of the comets of 1866 and 1366, first suggested by Professor H. A. Newton, is now unquestioned. The existence then of a meteoric swarm, moving in the same track, is not the only evidence of the original comet's partial dissolution. The comet of 1866 was invisible to the naked eye; that of 1366, seen under nearly similar circumstances, was a conspicuous object. The statement of the Chinese historian that "it appeared nearly as large as a tow measure,"[19] though somewhat indefinite, certainly justifies the conclusion that its magnitude has greatly diminished during the last 500 years. The meteors moving in the same orbit are doubtless the products of this gradual separation.

9. The repartition of Biela's comet in 1845, as well as the non-appearance of the two fragments in 1865 and 1872,[20] were referred to in a previous chapter.

The comet of Halley, if we may credit the descriptions given by ancient writers, has been decreasing in brilliancy from age to age. The same is true in regard to several others believed to be periodic. The comet of A.D. 1097 had a tail 50° long. At its return, in March, 1840, the length of its tail was only 5°. The third comet of 1790 and the first of 1825 are supposed, from the similarity of their elements, to be identical. Each perihelion passage occurred in May, yet the tail at the former appearance was 4° in length, at the latter but 2½°. Other instances might be specified of this apparent gradual dissolution. It would seem, indeed, extremely improbable that the particles driven off from comets in their approach to the sun, forming tails extending millions of miles from the principal mass, should again be collected around the same nuclei.

The fact, then, that meteors move in the same orbits with comets is but a consequence of that disruptive process so clearly indicated by the phenomena described. In this view of the subject, comets—even such as move in elliptic orbits—are not to be regarded as permanent members of the solar system. Their *débris* becomes gradually scattered around the orbit. Some parts of the nebulous ring will be more disturbed than others by planetary perturbation. Portions of such streams as nearly intersect the earth's path sometimes penetrate the atmosphere. Their rapid motion renders them luminous. If very minute, they are burnt up or dissipated without leaving any solid deposit; we then have the phenomena of *shooting-stars*. When, however, as is sometimes the case, they contain a considerable quantity of solid matter, they reach the earth's surface as *meteoric stones*.

---

[18] "Cometographia," p. 417.

[19] Williams' "Chinese Observations of Comets," p. 73.

[20] One of the parts was seen at Madras, India, on the mornings of December 2 and 3, 1872.

II.
METEORS.

CHAPTER VII.

METEORIC STONES.

Although numerous instances of the fall of aerolites had been recorded, some of them apparently well authenticated, the occurrence long appeared too marvelous and improbable to gain credence with scientific men. Such a shower of rocky fragments occurred, however, on the 26th of April, 1803, at L'Aigle, in France, as forever to dissipate all doubt on the subject. Similar displays since that time have been frequently witnessed;—indeed scarcely a year passes without the fall of meteoric stones in some part of the earth, either singly or in clusters. It would not comport with the design of the present treatise to give an extended list of these phenomena. The following account, however, includes the most important instances in which the fall of meteoric stones has been actually observed:

(1.) 1478 B.C.—According to the celebrated Parian chronicle, an aerolite, or *thunder-stone,* as it was called, fell in the island of Crete, about 1478 years before the Christian era. This is undoubtedly the most ancient stone-fall on record. Meteoric masses have been *found,* however, the fall of which *probably* occurred at an epoch still more ancient.

(2.) 1200 B.C.—A number of stones, which were anciently preserved in Orchomenos, a town of Bœotia, were said to have fallen from heaven about twelve centuries before our era.

(3.) 1168 B.C.—A mass of iron, as we learn from the Parian chronicle, was seen to descend upon Mount Ida, in Crete.

(4.) 654 B.C.—According to Livy, a number of meteoric stones fell on the Alban Hill, near Rome, about the year 654 B.C.

(5.) 616 B.C., *January* 14.—It is related in the Chinese annals that on the 14th of January, 616 B.C., a meteoric stone-fall broke several chariots and killed ten men.

(6.) 466 B.C.—A mass of rock, described as "of the size of two millstones," fell at Ægospotamos, in Thrace. An attempt to rediscover this meteoric mass, so celebrated in antiquity, was recently made, but without success. Notwithstanding this failure, Humboldt expressed the hope that, as such a body would be difficult to destroy, it may yet be found, "since the region in which it fell is now become so easy of access to European travelers."

(7.) 465 B.C.—The famous stone called the "Mother of the Gods," and which is described or alluded to by many ancient writers, was said to have fallen from the skies. The poet Pindar was seated on a hill at the time of its descent, and the meteorite struck the earth near his feet. The stone, as it fell, was *encircled by fire.* "It is said to have been of moderate dimensions, of a black hue, of an irregular, angular shape, and of a metallic aspect. An oracle had predicted that the Romans would continue to increase in prosperity if they were put in possession of this precious deposit; and Publius Scipio Nasico was accordingly deputed to Attalus, King of Pergamus, to obtain and receive the sacred idol, whose worship was instituted at Rome 204 years before the Christian era." — *Edinburgh Encyclopedia.*

(8.) A.D. 921. — An immense aerolite fell into the river (a branch of the Tiber) at Narni, in Italy. It projected three or four feet above the surface of the water.

(9.) 1492, *November* 7. — An aerolite, weighing 276 pounds, fell at Ensisheim, in Alsace, penetrating the earth to the depth of three feet. This stone, or the greater part of it, may still be seen at Ensisheim.

(10.) 1511, *September* 14. — At noon an almost total darkening of the heavens occurred at Crema. "During this midnight gloom," says a writer of that period, "unheard-of thunders, mingled with awful lightnings, resounded through the heavens.... On the plain of Crema, where never before was seen a stone the size of an egg, there fell pieces of rock of enormous dimensions and of immense weight. It is said that ten of these were found, weighing 100 pounds each." A monk was struck dead at Crema by one of these rocky fragments. This terrific display is said to have lasted two hours, and 1200 aerolites were subsequently found.

(11.) 1637, *November* 29. — A stone, weighing 54 pounds, fell on Mount Vaison, in Provence.

(12.) 1650, *March* 30. — A Franciscan monk was killed at Milan by the fall of a meteoric stone.

(13.) 1674. — Two Swedish sailors were killed on shipboard by the fall of an aerolite.

(14.) 1751, *May* 26. — Two meteoric masses, consisting almost wholly of iron, fell near Agram, the capital of Croatia. The larger fragment, which weighs 72 pounds, is now in Vienna.

(15.) 1790, *July* 24. — Between 9 and 10 o'clock at night a very large meteor was seen near Bordeaux, France. Over Barbotan a loud explosion was heard, which was followed by a shower of meteoric stones of various magnitudes.

(16.) 1794, *July*. — A fall of about a dozen aerolites occurred at Sienna, Tuscany.

(17.) 1795, *December* 13. — A large meteoric stone fell near Wold Cottage, in Yorkshire, England. "Several persons heard the report of an explosion in the air, followed by a hissing sound; and afterward felt a shock, as if a heavy body had fallen to the ground at a little distance from them. One of these, a plowman, saw a huge stone falling toward the earth, eight or nine yards from the place where he stood. It threw up the mould on every side; and after penetrating through the soil, lodged some inches deep in solid chalk-rock. Upon being raised, the stone was found to weigh 56 pounds. It fell in the afternoon of a mild, but hazy day, during which there was no thunder or lightning; and the noise of the explosion was heard through a considerable district." — *Milner's Gallery of Nature*, p. 134.

(18.) 1796, *February* 19.—A stone of 10 pounds' weight fell in Portugal.

(19.) 1803, *April* 26.—This remarkable shower was referred to on a previous page. At 1 o'clock P.M., the heavens being almost cloudless, a tremendous noise, like that of thunder, was heard, and at the same time an immense fire-ball was seen moving with great rapidity through the atmosphere. This was followed by a violent explosion, which lasted several minutes, and which was heard not only at L'Aigle, but in every direction around it to the distance of 70 miles. Immediately after, a great number of meteoric stones fell to the earth, generally penetrating to some distance beneath the surface. Nearly 3000 of these fragments were found and collected, the largest weighing about 17 pounds. The occurrence very naturally excited great attention. M. Biot, under the authority of the government, repaired to the place, collected the various facts in regard to the phenomenon, took the testimony of witnesses, etc., and finally embraced the results of his investigations in an elaborate memoir.

(20.) 1807, *December* 14.—A large meteor exploded over Weston, Connecticut. The height, direction, velocity and magnitude of this body were discussed by Dr. Bowditch in a memoir communicated to the American Academy of Arts and Sciences in 1815. The appearance of the meteor occurred about 6h. 15m. A.M.,—just after daybreak. Its apparent diameter was half that of the full moon; its time of flight, about 30 seconds. Within less than a minute from the time of its disappearance three distinct reports, like those of artillery, were heard over an area several miles in diameter. Each explosion was followed by the fall of meteoric stones. Unlike most aerolites, these bodies when first found were so soft as to be easily pulverized between the fingers. On exposure to the air, however, they gradually hardened. The weight of the largest fragment was 35 pounds.

(21.) 1859, *November* 15.—Between 9 and 10 o'clock in the morning an extraordinary meteor was seen in several of the New England States, New York, New Jersey, the District of Columbia, and Virginia. The apparent diameter of the head was nearly equal to that of the sun, and it had a train, notwithstanding the bright sunshine, several degrees in length. Its disappearance on the coast of the Atlantic was followed by a series of the most terrific explosions. It is believed to have descended into the water, probably into Delaware Bay. A highly interesting account of this meteor, by Professor Loomis, may be found in the *American Journal of Science and Arts* for January, 1860.

(22.) 1860, *May* 1.—About 20 minutes before 1 o'clock, P.M., a shower of meteoric stones fell in the southwest corner of Guernsey county, Ohio. Full accounts of the phenomena are given in *Silliman's Journal* for July, 1860, and January and July, 1861, by Professors E. B. Andrews, E. W. Evans, J. L. Smith, and D. W. Johnson. From these interesting papers we learn that the course of the meteor was about 40° west of north. Its visible track was over Washington and Noble counties, and the prolongation of its projection, on the earth's surface, passes directly through New Concord, in the southeast corner of Muskingum county. The meteor when first seen was about 40 miles from the earth's surface. The sky, at the time, was for the most part covered with clouds over northwestern Ohio, so that if any portion of the meteoric mass continued on its course it was invisible. The velocity of the meteor, in relation to the earth's surface, was from three to four miles per second; and hence its absolute velocity in the solar system must have been somewhat greater than that of the earth.

"At New Concord,[21] Muskingum county, where the meteoric stones fell, and in the immediate neighborhood, there were many distinct and loud reports heard. At New Concord there was first heard in the sky, a little southeast of the zenith, a loud detonation, which was compared to that of a cannon fired at the distance of half a mile. After an interval of ten seconds, another similar report. After two or three seconds another, and so on with diminishing intervals. Twenty-three distinct detonations were heard, after which the sounds became blended together and were compared to the rattling fire of an awkward squad of soldiers, and by others to the roar of a railway train. These sounds, with their reverberations, are thought to have continued for two minutes. The last sounds seemed to come from a point in the southeast 45° below the zenith. The result of this cannonading was the falling of a large number of stony meteorites upon an area of about 10 miles long by 3 wide. The sky was cloudy, but some of the stones were seen first as 'black specks', then as 'black birds', and finally falling to the ground. A few were picked up within 20 or 30 minutes. The warmest was no warmer than if it had lain on the ground exposed to the sun's rays. They penetrated the earth from two to three feet. The largest stone, which weighed 103 pounds, struck the earth at the foot of a large oak-tree, and, after cutting off two roots, one five inches in diameter, and grazing a third root, it descended two feet ten inches into hard clay. This stone was found resting under a root that was not cut off. This would seemingly imply that it entered the earth obliquely."

Over thirty of the stones which fell were discovered, while doubtless many, especially of the smaller, being deeply buried beneath the soil, entirely escaped observation. The weight of the largest ten was 418 pounds.

(23.) 1860, *July* 14.—About 2 o'clock P.M. on the 14th of July, 1860, a shower of aerolites fell at Dhurmsala, in India. The fall was attended by a tremendous detonation, which greatly terrified the inhabitants of the district. The natives, supposing the stones to have been thrown by some of their deities from the summit of the Himalayas, carried off many fragments to be kept as objects of religious veneration. Lord Canning and Mr. J. R. Saunders succeeded, however, in obtaining numerous specimens, which they forwarded to the British Museum and several European cabinets. They are earthy aerolites, of a specific gravity somewhat greater than that of granite.

(24.) 1864, *May* 14.—Early in the evening a very large and brilliant meteor was seen in France, from Paris to the Spanish border. At Montauban and in the vicinity loud explosions were heard, which were followed by showers of meteoric stones near the villages of Orgueil and Nohic. The principal facts in regard to the meteor are the following:

---

[21] New Concord is close to the Guernsey county line. Nearly all the stones fell in Guernsey.

Elevation when first seen, over	55 miles
Elevation at the time of its explosion	20 miles
Inclination of its path to the horizon	20° or 25°
Velocity per second, about	20 miles, or equal to that of the earth's orbital motion.

"This example," says Professor Newton, "affords the strongest proof that the detonating and stone-producing meteors are phenomena not essentially unlike."

(25.) 1868, *January* 30.—It is obviously a matter of much importance that the composition and general characteristics of aerolites, together with the phenomena attending their fall, should be carefully noted; as such facts have a direct bearing on the theory of their origin. In this regard the memoirs of Professors J. G. Galle, of Breslau, and G. vom Rath, of Bonn, on a meteoric fall which occurred at Pultusk, Poland, on the 30th of January, 1868, have more than ordinary interest. These memoirs establish the fact that the aerolites of the Pultusk shower *entered our atmosphere* as a swarm or cluster of distinct meteoric masses. It is shown, moreover, by Dr. Galle that this meteor-group had a proper motion when it entered the solar system of at least from 4½ to 7 miles per second.

The foregoing list contains but a small proportion of the meteoric stones whose fall has been actually observed. But, besides these, other masses have been found so closely similar in structure to aerolites whose descent has been witnessed, as to leave no doubt in regard to their origin. One of these is a mass of iron and nickel, weighing 1680 pounds, found by the traveler Pallas, in 1749, at Abakansk, in Siberia. This immense aerolite may be seen in the Imperial Museum at St. Petersburg. On the plain of Otumpa, in Buenos Ayres, is a meteoric mass 7½ feet in length, partly buried in the ground. Its estimated weight is about 16 tons. A specimen of this stone, weighing 1400 pounds, has been removed and deposited in one of the rooms of the British Museum. A similar block, of meteoric origin, weighing more than six tons, was discovered some years since in the province of Bahia, in Brazil.

General Remarks.

1. A Committee on Luminous Meteors was appointed several years since by the British Association for the Advancement of Science. This committee, consisting at present of James Glaisher, F.R.S., Robert P. Greg, F.R.S., Alexander S. Herschel, F.R.A.S., and Charles Brooke, F.R.S., report from year to year not only their own observations on aerolites, fire-balls, and falling stars, but also such facts bearing upon the subject as can be derived from other sources. An analysis of these reports justifies the conclusion that meteoric stone-falls, like star-showers, occur with greater frequency than usual on or about particular days. These epochs, established with more or less certainty, are the following:

(*a.*)	January	4th.
(*b.*)	January	16th.
(*c.*)	January	29th.

(*d.*)	February	10th.
(*e.*)	February	15th—18th.
(*f.*)	March	6th.
(*g.*)	March	12th.
(*h.*)	April	1st.
(*i.*)	April	10th—14th.
(*j.*)	May	8th—9th.
(*k.*)	May	13th—14th.
(*l.*)	May	17th—19th.
(*m.*)	June	3d.
(*n.*)	June	9th.
(*o.*)	June	12th.
(*p.*)	June	16th.
(*q.*)	July	3d—4th.
(*r.*)	July	14th—17th.
(*s.*)	August	5th—7th.
(*t.*)	August	11th.
(*u.*)	September	4th—10th.
(*v.*)	October	13th.
(*w.*)	November	5th.
(*x.*)	November	12th—13th.
(*y.*)	November	27th—30th.
(*z.*)	December	5th.
(*z´.*)	December	8th—14th.
(*z´´.*)	December	27th.

2. It is worthy of remark that no new elements have been found in meteoric stones. Humboldt, in his "Cosmos," called attention to this interesting fact. "I would ask," he remarks, "why the elementary substances that compose one group of cosmical bodies, or one planetary system, may not in a great measure be identical? Why should we not adopt this view, since we may conjecture that those planetary bodies, like all the larger or smaller agglomerated masses revolving round the sun, have been thrown off from the once far more expanded solar atmosphere, and have been formed from vaporous rings describing their orbits round the central body?"

3. But while aerolites contain no elements but such as are found in the earth's crust, the manner in which these elements are combined and arranged is so peculiar that a skillful mineralogist will readily distinguish them from terrestrial substances.

4. Of the eighteen or nineteen elements hitherto observed in meteoric stones, iron is found in the greatest abundance. The specific gravities vary from 1.94 to 7.901: the former being that of the stone of Alais; the latter that of the meteorite of Wayne county, Ohio, described by Professor J. L. Smith in *Silliman's Journal* for November, 1864, p. 385.

5. The average number of aerolitic falls in a year was estimated by Schreibers at 700. Baron Reichenbach, however, after a discussion of the data at hand, makes the number much larger. He regards the probable annual average for the entire surface of the earth as not less than 4500. This would give twelve daily falls. They are of every variety as to magnitude, from a weight of less than a single ounce to over fifteen tons. The baron even suspects the meteoric origin of large masses of dolerite which all former geologists had considered native to our planet.

6. An analysis of any extensive table of meteorites and fire-balls proves that a greater number of aerolitic falls have been observed during the months of June and July, when the earth is near its aphelion, than in December and January, when near its perihelion. It is found, however, that the reverse is true in regard to bolides, or fire-balls. These facts are susceptible of an obvious explanation. The fall of meteoric stones would be more likely to escape observation by night than by day, on account of the relatively small number of observers. But the days are shortest when the earth is in perihelion, and longest when in aphelion; the ratio of their lengths being nearly equal to that of the corresponding numbers of aerolitic falls. On the other hand, it is obvious that fire-balls, unless very large, would not be visible during the day. The *observed* number will therefore be greatest when the nights are longest; that is, when the earth is near its perihelion. This, it will be found, is precisely in accordance with observation.

CHAPTER VIII.

SHOOTING-STARS.—METEORS OF NOVEMBER 14.

Although shooting-stars have doubtless been observed in all ages of the world, it is only within the last half century that they have attracted the special attention of scientific men. A few efforts had been made to determine the height of such meteors, but the first general interest in the subject was excited by the brilliant meteoric display of November 13, 1833. This shower of fire can never be forgotten by those who witnessed it. The meteors were observed from the West Indies to British America, and from 60° to 100° west longitude from Greenwich. As early as 10 o'clock on the evening of the 12th shooting-stars were observed with unusual frequency; their motions being generally westward. Soon after midnight their numbers became so extraordinary as to attract the attention of all who happened to be in the open air. The meteors, however, became more and more numerous till 4, or half past 4, o'clock; and the fall did not entirely cease till ten minutes before sunrise. From 2 to 6 o'clock the numbers were so great as to defy all efforts at counting them; while their brilliancy was such that persons sleeping in rooms with uncurtained windows were aroused by their light. The meteors varied in apparent magnitude from the smallest visible points to fire-balls equaling the moon in diameter. Occasionally one of the larger class would separate into several parts, and in some instances a luminous train remained visible for three or four minutes. No sound whatever accompanied the display. It was noticed by many observers that all the meteors diverged from a point near the star *Gamma Leonis*; in other words, their paths if traced backward would intersect each other at a particular locality in the constellation Leo. In some parts of the country the inhabitants were completely terror-stricken by the magnificence of the display. In the afternoon of the day on which the shower occurred the writer met with an illiterate farmer who, after describing the phenomena as witnessed by himself, remarked that "the stars continued to fall till none were left," and added, "I am anxious to see how the heavens will appear this evening; I believe we shall see no more stars." A gentleman of South Carolina described the effect on the negroes of his plantation as follows:—"I was suddenly awakened by the most distressing cries that ever fell on my ears. Shrieks of horror and cries for mercy I could hear from most of the negroes of the three plantations, amounting in all to about 600 or 800. While earnestly listening for the cause I heard a faint voice near the door, calling my name. I arose, and, taking my sword, stood at the door. At this moment I heard the same voice still beseeching me to arise, and saying, 'O my God, the world is on fire!' I then opened the door, and it is difficult to say which excited me the most,—the awfulness of the scene, or the distressed cries of the negroes. Upwards of a hundred lay prostrate on the ground,—some speechless, and some with the bitterest cries, but with their hands raised, imploring God to save the world and them. The scene was truly awful; for never did rain fall much thicker than the meteors fell towards the earth; east, west, north, and south, it was the same."

At the time of this wonderful meteoric display Captain Hammond, of the ship *Restitution*, had just arrived at Salem, Massachusetts, where he observed the phenomenon from midnight till daylight. He recollected with astonishment that precisely one year before, viz., on the 13th of November, 1832, he had observed a similar appearance (although the meteors were less numerous) at Mocha, in Arabia. It was found, moreover, as a further and most remarkable coincidence, that an extraordinary fall of meteors had been witnessed on the 12th of November, 1799. This was seen and described by Andrew Ellicott, Esq., who was then at sea near Cape Florida. It was also observed by Humboldt and Bonpland, in Cumana, South America. Baron Humboldt's description of the shower is as follows:—"From half after two, the most extraordinary luminous meteors were seen toward the east. Thousands of bolides and falling stars succeeded each other during four hours. They filled a space in the sky extending from the true east 30° toward the north and south. In an amplitude of 60° the meteors were seen to rise above the horizon at E.N.E. and at E., describe arcs more or less extended, and fall toward the south, after having followed the direction of the meridian. Some of them attained a height of 40°, and all exceeded 25° or 30°. Mr. Bonpland relates, that from the beginning of the phenomenon there was not a space in the firmament equal in extent to three diameters of the moon, that was not filled at every instant with bolides and falling stars. The Guaiqueries in the Indian suburb came out and asserted that the firework had begun at one o'clock. The phenomenon ceased by degrees after four o'clock, and the bolides and falling stars became less frequent; but we still distinguished some toward the northeast a quarter of an hour after sunrise."

This wonderful correspondence of dates excited a very lively interest throughout the scientific world. It was inferred that a recurrence of the phenomenon might be expected, and accordingly arrangements were made for systematic observations on the 12th, 13th, and 14th of November. The periodicity of the shower was thus, in a very short time, placed wholly beyond question. The facts in regard to the phenomena of November 13, 1833, were collected and discussed by Olmsted, Twining, and other astronomers. The inquiry, however, very naturally arose whether any trace of the same meteoric group could be found in ancient times. To determine this question many old historical records were ransacked by the indefatigable scientist, Edward C. Herrick, in our own country, and by Arago, Quetelet, and others, in Europe. These examinations led to the discovery of ten undoubted returns of the November shower previous to that of 1799. The descriptions of these former meteoric falls are given by Professor H. A. Newton in the *American Journal of Science*, for May, 1864. They occurred in the years 902, 931, 934, 1002, 1101, 1202, 1366, 1533, 1602, and 1698. Historians represent the meteors of A.D. 902 as innumerable, and as moving like rain in all directions. The exhibition of 1202 was scarcely less magnificent. "On the last day of Muharrem," says a writer of that period, "stars shot hither and thither in the heavens, eastward and westward, and flew against one another like a scattering swarm of locusts, to the right and left; this phenomenon lasted until daybreak; people were thrown into consternation, and cried to God the Most High with confused clamor." The shower of 1366 is thus described in a Portuguese chronicle, quoted by Humboldt: "In the year 1366, twenty-two days of the month of October being past, three months before the death of the king, Don Pedro (of Portugal), there was in the heavens a movement of stars such as men never before saw or heard of. At midnight, and for some time after, all the stars moved from the east to the west; and after being collected together, they began to move, some in one direction and others in another. And afterward they fell from the sky in such numbers, and so thickly together, that as they descended low in the air they seemed large and fiery, and the sky and the air seemed to be in flames, and even the earth appeared as if ready to take fire. That portion of the sky where there were no stars seemed to be divided into many parts, and this lasted for a long time."

The Showers of 1866-9.

The fact that all great displays of the November meteors have taken place at intervals of 33 or 34 years, or some multiple of that period, had led to a general expectation of a brilliant shower in 1866. In this country, however, the public curiosity was much disappointed.[22] The numbers seen were greater than on ordinary nights, but not such as would have attracted any special attention. The greatest number recorded at any one station was seen at New Haven by Professor Newton. On the night of the 12th 694 were counted in five hours and twenty minutes, and on the following night, 881 in five hours. A more brilliant display was, however, witnessed in Europe. Meteors began to appear in unusual frequency about 11 o'clock on the night of the 13th, and their numbers continued to increase with great rapidity for more than two hours; the maximum being reached a little after 1 o'clock. A writer in Edinburgh, Scotland, thus describes the phenomenon as observed at that city:—"Standing on the Calton Hill, and looking westward,—with the observatory shutting out the lights of Princes Street,—it was easy for the eye to delude the imagination into fancying some distant enemy bombarding Edinburgh Castle from long range; and the occasional cessation of the shower for a few seconds, only to break out again with more numerous and more brilliant drops of fire, served to countenance this fancy. Again, turning eastward, it was possible now and then to catch broken glimpses of the train of one of the meteors through the grim dark pillars of that ruin of most successful manufacture, the National Monument; and in fact from no point in or out of the city was it possible to watch the strange rain of stars, pervading as it did all points of the heavens, without pleased interest and a kindling of the imagination, and often a touch of deeper feeling that bordered on awe." At London about 1 o'clock a single observer counted 200 in two minutes. The whole number seen at Greenwich was 8485. The shower was also observed in different countries on the continent.

---

[22] The first indication of the approaching shower was the appearance of meteors in unusual numbers at Malta, on the 13th of November, 1864. In 1865, as observed at Greenwich and other stations, they were still more numerous.

In 1867 the display was generally observed throughout the United States. From the able and interesting reports of Commodore Sands and Professors Newcomb, Harkness, and Eastman, we derive the following facts in regard to the shower as seen at Washington, D. C.:

Commencement	$1h.$	$0m.$	A.M.	Nov. 14.
Maximum	4	20	A.M.	Nov. 14.
End	5	0	A.M.	Nov. 14.

Number of meteors per hour at maximum 3000

Mean height on first appearance 75 miles.

Mean height on disappearance 55 miles.

Position of radiant, R. A. 151°, Decl. 22½°.

The shower of 1868 was in some respects quite remarkable, though the number of meteors was less than in 1866 or 1867. At New Haven the fall commenced about midnight, and from 2 o'clock till daybreak over 5000 meteors were counted. The time of maximum could not be accurately determined, as no decrease in the numbers was observable till dawn. The display was also witnessed in England and in Cape Colony, South Africa. The times of maxima in these countries differed so materially as to indicate a decided stratification of the meteoric stream. The entire depth, moreover, where crossed by the earth in 1868, was much greater than at the part traversed either in 1866 or 1867.

In 1869 the shower was observed at Port Saïd, Lower Egypt, by G. L. Tupman, Esq.; in Florida, U. S., by Commander William Gibson, U.S.N.; and at Santa Barbara, California, by Mr. G. Davidson and Mrs. E. Davidson. The first observed 112 meteors in 1h. 54m., from 2h. 30m. to 4h. 24m., Alexandria mean time; the numbers during this interval being nearly equal, though slightly decreasing. Throughout the morning (November 14) the sky was only partly clear. The two observers at Santa Barbara saw 556 in 2h. 25m., ending at 3h. 43m. A.M. In Florida also the display was quite brilliant, though inferior to that of 1868. It should be remarked that the morning in many parts of the United States was cloudy. No considerable number of the meteors of this stream has been observed in any part of the world since 1869.

DISCUSSION OF THE PHENOMENA.

Since the memorable display of November 13, 1833, the phenomena of shooting-stars have been observed and discussed with a very lively interest. Among the first laborers in this department of research the names of Olmsted, Herrick, and Twining must ever hold a conspicuous place. The fact that the position of the radiant point did not change with the earth's rotation at once placed the cosmical origin of the meteors wholly beyond question. The theory of a ring of nebulous matter revolving round the sun in an elliptic orbit—a theory somewhat different from that proposed by Olmsted—was found to afford a simple and satisfactory explanation of the phenomena. This hypothesis of an eccentric stream of meteors intersecting the earth's orbit was adopted by Humboldt, Arago, and others, shortly after the occurrence of the meteoric shower of 1833.

A few years previous to the display of 1866 it was shown by Professor Newton, of Yale College, that the distribution of meteoric matter around the ring or orbit is far from uniform; that the motion is retrograde; that the node of the orbit has an annual forward motion of 102″.6 with respect to the equinox, or of 52″.4 with respect to the fixed stars; that the periodic time must be limited to five accurately determined periods, viz.: 180.05 days, 185.54 days, 354.62 days, 376.5 days, or 33.25 years; and that the inclination of the orbit to the ecliptic is about 17°. Professor Newton, for reasons assigned, regarded the third period named as the most probable. He remarked, however, that by computing the secular motion of the node for each periodic time, and comparing the result with the known precession, it was possible to determine which of the five periods is the correct one.

For the application of this crucial test,—a problem of more than ordinary interest,—we are indebted to Professor J. C. Adams, of Cambridge, England. By an elegant analysis it was first shown that for either of the first four periods designated by Professor Newton, the annual motion of the node, resulting from planetary perturbation, would be considerably less than one half of the observed motion. It only remained, therefore, to examine whether the period of 33¼ years would give a motion of the node corresponding with observation. Professor Adams found that in this time the longitude of the node is increased 20′ by the action of Jupiter, 7′ by the action of Saturn, and 1′ by that of Uranus. The effect of the other planets is scarcely perceptible. The *calculated* motion in 33¼ years is therefore 28′. The *observed* motion in the same time, according to Professor Newton, as previously stated, is 29′. This remarkable accordance was at once accepted by astronomers as satisfactory evidence that the period is about 33.25 years.

Having determined the periodic time, the mean distance, or semi-axis major, is found by Kepler's third law to be 10.34. The aphelion is consequently situated at a comparatively short distance beyond the orbit of Uranus. The orbit is represented in Fig. 4.

It was stated at the close of Chapter VI. that shooting-stars are the dissevered fragments of cometic matter, which, penetrating our atmosphere, are rendered luminous by the resistance so encountered. The discovery that comets and meteors are actually moving in the same orbits was first announced by Signor Schiaparelli in 1867. The coincidence of the orbits of Tempel's comet[23] as computed by Dr. Oppolzer, and the meteors of November 14 as determined by Schiaparelli, is too close to be regarded as merely accidental. These elements are as follows:

	Nov. Meteors.	Tempel's Comet.
Perihelion passage	Nov. 10.092, 1866.	Jan. 11.160, 1866.
Passage of descending node	Nov. 13.576,	
Longitude of perihelion	56° 26′	60° 28′
Longitude of ascending node	231° 28′	231° 26′

---

[23] See page 11.

Inclination	17° 44′	17° 18′
Perihelion distance	0.9873	0.9765
Eccentricity	0.9046	0.9054
Semi-major axis	10.3400	10.3240
Periodic time	33.2500 $y.$	33.1760 $y.$
Motion	Retrograde.	Retrograde.

Fig. 4.

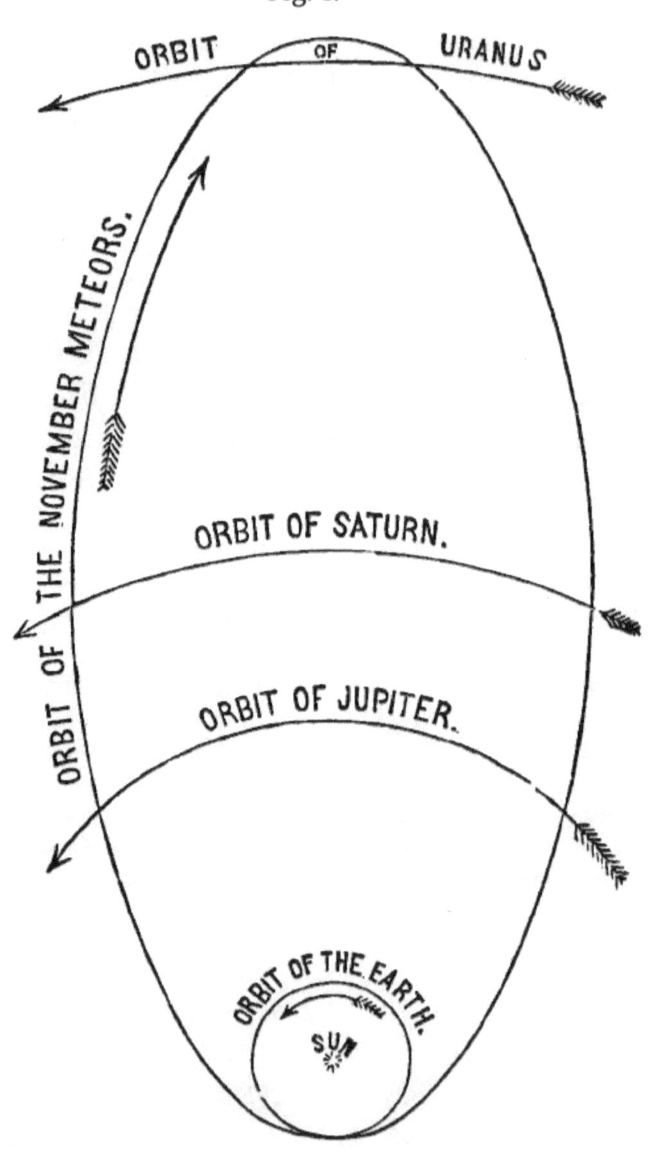

The fact is thus obvious that the meteors of November 14 are the products of the comet's gradual dissolution. It has been stated that the comets of 1366 and 1866 are probably identical. The interval indicates a period of 33.283 years—greater by 39 days than that found by Oppolzer. With this value of the periodic time and the known secular variation of the node it is found that the comet and Uranus were in close proximity about the beginning of the year 547 B.C. It is therefore not improbable that the former was then thrown into its present orbit by the attraction of the latter. The celebrated Leverrier designated the year 126 of our era as the probable epoch of the comet's entrance into our system. This date, however, is incompatible with the period here adopted. It is worthy of remark, moreover, as bearing on this question, that the extension of the cluster in the tenth century, as indicated by the showers of 902, 931, and 934, was too great to have been effected in so short a period as 800 years.

With the period of 33.283 years it is easy to find that the comet will make a near approach to the earth about the 16th or 17th of November, 1965, and to Uranus in 1983. At one of these epochs the cometary orbit will probably undergo considerable transformation.

We have seen that the comet of 1866, and also the meteoroids following in its path, have their perihelion at the orbit of the earth, and their aphelion at the orbit of Uranus. Both planets, therefore, at each encounter with the current not only appropriate a portion of the meteoric matter, but entirely change the orbits of many meteoroids. In regard to the devastation produced by the earth in passing through the cluster, it is sufficient to state that, according to Weiss, the meteor orbits resulting from the disturbance will have all possible periods from 21 months to 390 years. It may be regarded, therefore, as evidence of the recent[24] introduction of this meteor-stream into the solar system that the comet of 1866, which constitutes a part of the cluster, has not been deflected from the meteoric orbit by either the earth or Uranus.

---

[24] Recent in comparison with the origin of the August meteors, which constitute a continuous ring.

CHAPTER IX.

OTHER METEORIC STREAMS.

The Meteors of August 7-11.—Muschenbroek, in his "Introduction to Natural Philosophy," published in 1762, stated as the result of his own observations that shooting-stars are more abundant in August than in any other part of the year. The fact, however, that a maximum occurs on the 9th or 10th of the month was first shown by Quetelet in 1835. Since that time the shower has been regularly observed both in Europe and America; the number of meteors at the maximum sometimes amounting to 160 per hour. Their tracks when produced backward intersect each other at a particular point in the constellation Perseus.

Of the 315 meteoric displays given in Quetelet's catalogue, 63 belong to the August epoch. Their dates up to the commencement of the present century are as follows:

1.	A.D.	811,	July	25th.
2.		820,	July	25th-30th.
3.		824,	July	26th-28th.
4.		830,	July	26th.
5.		833,	July	27th.
6.		835,	July	26th.
7.		841,	July	25th-30th.
8.		924,	July	27th-30th.
9.		925,	July	27th-30th.
10.		926,	July	27th-30th.
11.		933,	July	25th-30th.
12.		1243,	Aug.	2d.
13.		1451,	Aug.	7th.
14.		1709,	Aug.	8th.
15.		1779,	Aug.	9th-10th.
16.		1781,	Aug.	8th.
17.		1784,	Aug.	6th-9th.
18.		1789,	Aug.	10th.

19.	1798,	Aug.	9th.
20.	1799,	Aug.	9th-10th.
21.	1800,	Aug.	10th.

As the earth is about five days in crossing the ring, its breadth in some parts cannot be less than 8,000,000 miles.

In 1866 Professor Schiaparelli, on computing the orbit of this meteoric stream, noticed the remarkable agreement of its elements with those of Swift's or Tuttle's comet[25] (1862, III.), as computed by Dr. Oppolzer. These coincidences are exhibited in the following table:

	Meteors of August 10.	Comet III. of 1862.
Longitude of perihelion	343° 38′	344° 41′
Ascending node	138° 16′	137° 27′
Inclination	63° 3′	66° 25′
Perihelion distance	0.9643	0.9626.
Period	105 years (?)	121.5 years.
Motion	Retrograde.	Retrograde.

It appears, therefore, that the third comet of 1862 is a part of the meteoric stream whose orbit is crossed by the earth on the 10th of August.

The characteristics of different meteor-zones afford interesting indications in regard to their relative age, the magnitude and composition of their corpuscles, etc. Thus, if we compare the streams of August 10 and November 14, we shall find that the former probably entered our system at a comparatively remote epoch. We have seen that at each return to perihelion the meteoric cluster is extended over a greater arc of its orbit. Now, Tuttle's comet and the August meteors undoubtedly constituted a single group previous to their entering the solar domain. It is evident, however, from the annual return of the shower during the last 90 years, that the ring is at present nearly if not quite continuous. That the meteoric mass had completed many revolutions before the ninth century of our era is manifest from the frequent showers observed between the years 811 and 841. At the same time, the long interval of 83 years between the last observed display in the ninth century, and the first in the tenth, seems to indicate the existence of a wide chasm in the ring no more than a thousand years since.

Neither the period of the meteors nor that of the comet can yet be regarded as accurately ascertained. The latter, however, in all probability, exceeds the former by several years. Now, at each passage of the earth through the elliptic stream, those meteoroids nearest the disturbing body must be thrown into orbits differing more or less from that of the primitive group. In like manner the near approach of the *comet* to the earth at an ancient epoch may account for the lengthening of its periodic time.

---

[25] Mr. Swift, of Marathon, N. Y., had two or three days priority in the discovery of this comet, but unfortunately delayed his announcement of the fact.

The Meteors of November 27.

Professor Schiaparelli's brilliant discovery of the relation between comets and meteors may now be ranked with the established truths of astronomy. His hypothesis, however, in regard to the *origin* of meteoric streams has not been generally accepted. Comets and meteors, according to his theory, are derived from cosmical clouds existing in great numbers in stellar space. These nebulæ, in consequence of their own motion or that of the sun, are drawn towards the centre of our system. By the unequal influence of the sun's attraction on different parts, such clouds are transformed into currents of great length before reaching the limits of the planetary system. Shooting-stars, fire-balls, aerolites, and comets being all of the same nature, differing merely in size, sometimes fall towards the sun as parts of the same current.

The views of Dr. Weiss, of Vienna, differ from those of Schiaparelli, in that he regards comets as the original bodies by whose disintegration meteor-streams are gradually formed.[26] "Cosmical clouds," he remarks, "undoubtedly appear in the universe, but only of such density that in most cases they possess sufficient coherence to withstand the destructive operation of the sun's attraction, not only up to the boundaries of our solar system, but even within it. Such cosmical clouds will always appear to us as comets when they pass near enough to the earth to become visible. Approaching the sun, the comet undergoes great physical changes, which finally affect the stability of its structure: it can no longer hold together: parts of it take independent orbits around the sun, having great resemblance to the orbit of the parent comet. With periodical comets, this process is repeated at each successive approach to the sun. Gradually the products of disintegration are distributed along the comet's orbit, and if the earth's orbit cuts this, the phenomenon of shooting-stars is produced."

These views of the distinguished astronomer of Vienna are confirmed by the star-shower of November 27, 1872. That the orbits of the earth and Biela's comet intersect at the point passed by the former about the last of November, and that in 1845 the comet separated into two visible parts, has been stated in a previous chapter. The comet's non-appearance in December, 1865, and in September, 1872, was regarded by astronomers as presumptive evidence of its progressive dissolution. A meteoric shower, resulting from the earth's collision with the cometary *débris*, was accordingly expected about the 27th of November.

The first indication of the approaching display appeared on the evening of November 24, when meteors in unusual numbers were observed by Professor Newton, at New Haven, Connecticut. On Wednesday evening, the 27th, from the close of twilight till 8 o'clock, a decided shower of shooting-stars was noticed in various parts of the United States. At Greencastle, Indiana, Professor Joseph Tingley counted 110 meteors in 40 minutes, and at Princeton, in the same State, Mr. D. Eckley Hunter counted 70 in 80 minutes. The numbers seen at New Haven were considerably greater. The fact that the display commenced before daylight had entirely closed seemed to indicate that only the termination of the shower had

[26] *Astr. Nach.*, Nos. 1710, 1711. For a fuller statement of Schiaparelli's theory, see Silliman's Journal for May, 1867.

been observed in this country. Accordingly the display was soon found to have been witnessed from 60° E. to 90° W. of Greenwich, or through 150° of longitude. In England the first bolide of the swarm was seen by M. M. Brinkley, at 3 o'clock, P.M., in full daylight. The meteors were most numerous in the southern part of the continent, particularly in Italy. At the Observatory of Breslau, according to M. Faye, 3000 were seen from 6h. 30m. to 7h. 50m. Dr. Heis reported that at Münster 2500 per hour were counted by two observers. At Naples, Signor Gasparis observed two meteors per second. At Turin, M. Denza, Director of the Observatory, reported 33,400 in 6h. 30m.; many of various and delicate colors, and followed by long and brilliant trains. At some points the numbers were so great that an accurate enumeration was wholly impossible. In short, the display was decidedly the most brilliant that has occurred since that of November 13, 1833.

But some of the most interesting circumstances in connection with the phenomena of November 27, 1872, remain to be detailed. Astronomers without exception regarded the display as due to the earth's passage through the *débris* following in the path of Biela's comet. In accordance with this view Dr. Klinkerfues, of Gottingen, concluded that the comet itself, or rather its largest portion, ought to be found in the region of the heavens nearly opposite to that from which the meteoroids appeared to radiate.[27] As this point in the southern hemisphere could not be observed in Europe, he conceived the happy idea of detecting the fugitive *by means of the electric telegraph*. The following was accordingly dispatched to Mr. Pogson, Director of the Government Observatory at Madras, in Southern India: *"Biela touched earth on 27th; search near Theta Centauri."* The first two mornings after the receipt of this dispatch were cloudy at Madras. On the third, however, the cometary fragment was found, and its motion accurately measured. The observer described it as circular and rather bright, with no traces of a tail. But one fragment could be detected. On the next morning, December 3, the comet was again observed. Its diameter had sensibly increased; it had a bright nucleus, and still presented a circular aspect. A faint tail was also noticed, equal in length to one-fourth of the moon's apparent diameter. The following mornings being again cloudy, no further observations could be obtained. This cometary mass will be in close proximity to the earth about the last of November, 1892. Another brilliant meteoric shower may therefore be expected at that epoch.

The Meteors of April 20.

Meteoric showers have occurred about the 20th of April in the following years:

$$
\begin{array}{rr}
\text{B.C.} & 687 \\
& 15 \\
\text{A.D.} & 582 \\
& 1093 \\
& 1094 \\
& 1095 \\
& 1096 \\
\end{array}
$$

---

[27] The radiant of the Biela meteors is near *Gamma Andromedæ*.

$$\left. \begin{array}{l} 1122 \\ 1123 \end{array} \right\}$$

$$1803$$

The probability that these meteors are derived from a ring which intersects the earth's orbit, was first suggested by Arago in 1836. A comparison of dates led Herrick to designate 27 years as the probable period of the cluster. In the *Astronomische Nachrichten*, No. 1632, Dr. Weiss called attention to the fact that the orbit of the first comet of 1861 very nearly intersects that of the earth, in longitude 210° — the point passed by the latter at the epoch of the April meteoric shower. A relation between the meteors and the comet, indicating an approximate equality of periods, was thus suggested as probable. But the comet, according to Oppolzer, does not complete a revolution in less than 415 years. If, therefore, the meteoric period is nearly the same, the known dates of star-showers indicate a diffusion of meteoroids around one half of the orbit previous to the display of the year 15 B.C. No subsequent perturbation, then, of a particular *part* could sensibly effect the general orbit of the stream. The infrequency of the display renders, therefore, the hypothesis of a long period extremely improbable.

The entire interval between 687 B.C. and A.D. 1803 is 2490 years, or 92 periods of 27.0652 years; and the known dates are all satisfied by the following scheme:

B.C.	687 to B.C.	15 ... 672	years	=	25 periods of	26.8800 *y*.	each.
	15 to A.D.	582 ... 597	years	=	22 periods of	27.1363 *y*.	each.
A.D.	582 to	1095 ... 513	years	=	19 periods of	27.0000 *y*.	each.
	1095 to	1122 ... 27	years	=	1 periods of	27.0000 *y*.	each.
	1122 to	1803 ... 681	years	=	25 periods of	27.2400 *y*.	each.

With a period of 27 years, the perihelion being interior to the earth's orbit, the aphelion distance of the meteors would be very nearly equal to the distance of Uranus. The next shower, if the assumed period be correct, ought to occur about 1884. It is worthy of remark that near the time of the last (hypothetical) return Mr. Du Chaillu witnessed the meteors of this epoch, in considerable numbers, in the interior of Africa.

The Meteors of December 12.

Meteoric showers have occurred about the 12th of December in the following years:

1. A.D. 901. "The whole hemisphere was filled with those meteors called falling-stars from midnight till morning, to the great surprise of the beholders in Egypt."

2. In 930 a remarkable shower of falling stars was observed in China.

3. Extraordinary meteoric phenomena were observed at Zurich at the same epoch in 1571.

4. On the night of the 11th and 12th of December, 1833, a great number of

shooting-stars were seen at Parma. At the maximum as many as ten were visible at the same time.

5. (Doubtful.) 1861, 1862, and 1863. Maximum probably in 1862. The meteors at this return were far from being comparable in numbers with the ancient displays. The shower, however, was distinctly observed. R. P. Greg, Esq., of Manchester, England, says the period of December 12, 1862, was "exceedingly well defined."

These dates indicate a period of about 29⅛ years. Thus:

$$901 \text{ to } 930 \quad 1 \text{ period of } 29.000 \text{ years.}$$
$$930 \text{ to } 1571 \quad 22 \text{ period of } 29.136 \text{ years.}$$
$$1571 \text{ to } 1833 \quad 9 \text{ period of } 29.111 \text{ years.}$$
$$1833 \text{ to } 1862 \quad 1 \text{ period of } 29.000 \text{ years.}$$

Meteors of October 16-20.

Meteoric showers were observed from the 16th to the 20th of October in the years 288, 1436, 1439, 1743, and 1798. These dates render it somewhat probable that the period is about 27½ years. Thus:

$$\text{A.D.} \quad 288 \text{ to } 1439 \quad 42 \text{ periods of } 27.405 \text{ years each.}$$
$$1439 \text{ to } 1743 \quad 11 \text{ periods of } 27.636 \text{ years each.}$$
$$1743 \text{ to } 1798 \quad 2 \text{ periods of } 27.500 \text{ years each.}$$

If these periods are correct, it is a remarkable coincidence that the aphelion distances of the meteoric rings of April 20, October 18, November 14, and December 12, as well as those of the comets 1866 I., and 1867 I., are all nearly equal to the mean distance of Uranus.

The Meteors of April 30, May 1.

Professor Schiaparelli, in his list of meteoric showers whose radiant points are derived from observations made in Italy during the years 1868, 1869, and 1870, describes one as occurring on April 30 and May 1; the radiant being in the Northern Crown. The same shower has also been recognized by R. P. Greg, F.R.S., of Manchester, England. This meteor-stream, it is now proposed to show, is probably derived from one much more conspicuous in ancient times.

In Quetelet's "Physique du Globe" we find meteoric displays of the following dates. In each case the corresponding day for 1870 is also given,[28] in order to exhibit the close agreement of the epochs:

1. A.D.   401, April   9th; corresponding to April 29th, for 1870.
2.   538, April   6th; corresponding to April 25th, for 1870.
3.   839, April 17th; corresponding to May 1st,   for 1870.
4.   927, April 17th; corresponding to April 30th, for 1870.

[28] Making proper allowance for the precession of the equinoxes.

5.   934, April 18th; corresponding to May 1st,   for 1870.

6.   1009, April 16th; corresponding to April 28th, for 1870.

The epochs of 927 and 934 suggest as probable the short period of 7 years. It is found accordingly that the entire interval of 608 years—from 401 to 1009—is equal to 89 mean periods of 6.8315 years each. With this approximate value the six dates are all represented as follows:

From A.D. 401 to A.D.   538, 20 periods of 6.85 years.

           538 to   839, 44 periods of 6.84 years.

           839 to   927, 13 periods of 6.77 years.

           927 to   934,   1 periods of 7.00 years.

           934 to   1009, 11 periods of 6.82 years.

This period nearly corresponds to those of several comets whose aphelion distances are somewhat greater than the mean distance of Jupiter. So long as the cluster occupied but a small arc of the orbit the displays would evidently be separated by considerable intervals. The comparative paucity of meteors in modern times may be explained by the fact that the ring has been subject to frequent perturbations by Jupiter.

Groups in which the Meteoroids are sparsely scattered.

By the labors of Heis, Greg, Herschel, Schiaparelli, and others, the radiants of more than fifty sparsely strewn meteor-systems have been determined. Of these the following, which are well defined, seem worthy of special study:

DATE.	POSITION OF RADIANT.	
	R. A.	N. Decl.
January 1-4	234°	51°
January 18	232°	36°
April 25	142°	53°

The orbits and periods, except in the few cases previously considered, are entirely unknown. Some of the observed clusters are probably the *débris* of ancient comets whose aphelia were in the vicinity of Jupiter's orbit.

CHAPTER X.

THE ORIGIN OF COMETS AND METEORS.

The fact that comets and meteors, or at least a large proportion of such bodies, have entered the solar system from stellar space, is now admitted by all astronomers. The question, however, in regard to the origin and nature of these cosmical clouds still remains undecided. The theory that they consist of matter expelled with great velocity from the fixed stars appears to harmonize the greatest number of facts, and is accordingly entitled to respectful consideration. The evidence by which it is sustained may be briefly stated as follows:

1. The observations of Zollner, Respighi, and others, have indicated the operation of stupendous eruptive forces beneath the solar surface. The rose-colored prominences, which Janssen and Lockyer have shown to be masses of incandescent hydrogen, are regarded by Professor Respighi as phenomena of eruption. "They are the seat of movements of which no terrestrial phenomenon can afford any idea; masses of matter, the volume of which is many hundred times greater than that of the earth, completely changing their position and form in the space of a few minutes." The nature of this eruptive force is not understood. We may assume, however, that it was in active operation long before the sun had contracted to its present dimensions.

2. With an initial velocity of projection equal to 380 miles per second, the matter thrown off from the sun would be carried beyond the limits of the solar system, never to return. With velocities somewhat less, it would be transported to distances corresponding to those of the aphelia of the periodic comets.

3. On the 7th of September, 1871, Professor Young, of Dartmouth College,[29] witnessed an extraordinary explosion on the sun's surface. The observer, with his telescope, followed the expelled matter to an elevation of over 200,000 miles. The mean velocity between the altitudes of 100,000 and 200,000 miles was 166 miles per second. This rate of motion *in vacuo* would indicate an initial velocity of about 260 miles per second. But the sun is surrounded by an extensive atmosphere, whose resistance must have greatly retarded the velocity of the outrush before reaching the height of 100,000 miles. The original velocity of these hydrogen clouds was therefore sufficient, in all probability, to have carried them, if unresisted, beyond the solar domain. Solid or dense matter propelled with equal force would doubtless have been driven off never to return.[30]

[29] Boston Journal of Chemistry, November, 1871.

[30] See Mr. Proctor's interesting discussion of this subject in the Monthly Notices of the R.A.S., vol. xxxii.

4. This eruptive force, whatever be its nature, is probably common to the sun and the so-called fixed stars. If so, the dispersed fragments of ejected matter ought to be found in the spaces intervening between sidereal systems. Accordingly, the phenomena of comets and meteors have demonstrated the existence of such matter, widely diffused, in the portions of space through which the solar system is moving.

5. According to Mr. Sorby the microscopic structure of the aerolites he has examined points evidently to the fact that they have been at one time in a state of fusion from intense heat,—a fact in striking harmony with this theory of their origin.

6. The velocity with which some meteoric bodies have entered the atmosphere has been greater than that which would have been acquired by simply falling toward the sun from any distance, however great. On the theory of their sidereal origin, this excess of velocity has been dependent on the primitive force of expulsion. The shower of aerolites which fell at Pultusk, Poland, on the 30th of January, 1868,[31] is not only a remarkable illustration of the fact here stated, but also of another which may be accounted for by the same theory, viz.: that meteoric bodies sometimes enter the solar system in groups or clusters.

7. A striking argument in favor of this theory may be derived from the researches of the late Professor Graham, considered in connection with those of Dr. Huggins and other eminent spectroscopists. Professor Graham found large quantities of hydrogen confined in the pores or cavities of certain meteoric masses. Now, the spectroscope has shown that the sun's rose-colored prominences consist of immense volumes of incandescent hydrogen; that the same element exists in great abundance in many of the fixed stars, and even in certain nebulæ; and that the star in the Northern Crown, whose sudden outburst in 1866 so astonished the scientific world, afforded decided indications of its presence.

THE END.

---

[31] See Chapter VII.

Lector House believes that a society develops through a two-fold approach of continuous learning and adaptation, which is derived from the study of classic literary works spread across the historic timeline of literature records. Therefore, we aim at reviving, repairing and redeveloping all those inaccessible or damaged but historically as well as culturally important literature across subjects so that the future generations may have an opportunity to study and learn from past works to embark upon a journey of creating a better future.

This book is a result of an effort made by Lector House towards making a contribution to the preservation and repair of original ancient works which might hold historical significance to the approach of continuous learning across subjects.

HAPPY READING & LEARNING!

LECTOR HOUSE LLP
E-MAIL: lectorpublishing@gmail.com

www.ingramcontent.com/pod-product-compliance
Lightning Source LLC
Chambersburg PA
CBHW021356160726
47994CB00007B/2977